BestMasters

Springer awards „BestMasters" to the best master's theses which have been completed at renowned Universities in Germany, Austria, and Switzerland.

The studies received highest marks and were recommended for publication by supervisors. They address current issues from various fields of research in natural sciences, psychology, technology, and economics.

The series addresses practitioners as well as scientists and, in particular, offers guidance for early stage researchers.

Janne Marie Soetbeer

Dynamical Decoupling in Distance Measurements by Double Electron- Electron Resonance

Janne Marie Soetbeer
Zurich, Switzerland

BestMasters
ISBN 978-3-658-14669-6   ISBN 978-3-658-14670-2 (eBook)
DOI 10.1007/978-3-658-14670-2

Library of Congress Control Number: 2016943852

Springer Spektrum
© Springer Fachmedien Wiesbaden 2016

Printed on acid-free paper

This Springer Spektrum imprint is published by Springer Nature
The registered company is Springer Fachmedien Wiesbaden GmbH

"I am not aware of any other field of science outside of magnetic resonance that offers so much freedom and opportunities for a creative mind to invent and explore new experimental schemes (...)"

Richard R. Ernst, Nobel Lecture, December 9, 1992

Acknowledgements

I would like to thank Yevhen Polyhach for the supervision of this master thesis. Thanks goes to Kamila Guerin for showing me the spin labelling technique for the T4 Lysozyme sample. This thesis would have not been possible without Andreas Dounas work on the optimisation of adiabatic inversion pulse envelopes. I am very much obliged to also thank Andrin Doll for his help at setting up all AWG measurements at both X and Q band. Finally, I would like to thank Professor Gunnar Jeschke for the opportunity to carry out my master thesis in his group and for all the support and discussion related to this thesis.

Abstract

The relaxation behaviour of model systems mimicking different spin-labelled electron spin environments have been studied including TEMPOL, spin-labelled T4 Lysozyme, spin-labelled WALP23 and rigid spin-labelled molecules. Carr-Purcell (CP) and Uhrig pulse sequences for a fixed total sequence time were used to record echo relaxation traces to find the optimal dynamical decoupling (DD) scheme which allows for efficient reduction of electron spin coherence losses for a given relaxation behaviour. Addition of π refocusing pulses led to stronger nuclear modulation of the recorded relaxation traces and to signal contributions from both refocused and stimulated echos at the acquisition position. The latter complication was alleviated by small time shifts introduced into the pulse sequence.

The relaxation behaviour was characterised by phase relaxation time T_m and stretch parameter ξ which were shown to both depend on the spin environment and the dominating relaxation mechanism. Deviations from literature values could be explained by the inherent nature of the relaxation measurement. DD schemes proofed most successful for $\xi > 1$ as expected from theory. Overall, Uhrig schemes allowed for stronger suppression of coherence losses for interpulse delays at which the magnetisation in CP-type set ups had already decayed.

Based on the relaxation characterisation the DD schemes were extended to corresponding DD-double electron electron resonance (DEER) experiments by addition of pump pulses for a rigid biradical model system in d_8-OTP and T4 Lysozyme in H_2O-d_8-Glycerol mixture. CP-type DD observer pulse sequences of order $n = 1, 2$ and 4 were used to record DEER traces. The promising relaxation study results for Uhrig-type DD schemes did not translate into a superior Uhrig-DEER experiment relative to a CP derived set up for different reasons at X and Q band.

Signal artifacts related to m times pump spin inversion with $1 \leq m < n$ in DD-DEER experiments of order $n = 2$ or 4 were strongly reduced by increasing inversion efficiency based on pump pulse optimisation. Linear-chirp and asymmetric hyperbolic secant pulses were tested with optimal pulse parameters obtained from pulse simulation including resonator profile compensation. Measurements at Q band showed pump pulse underperformance with respect to inversion efficiencies expected from theory. This resulted in strong contributions from signal artifacts in the corresponding DD-DEER traces. Nearly full suppression of the latter were achieved at X band with sacrifices in terms of sensitivity relative to Q band.

X-band order $n = 2$ DD-DEER traces for the low concentration rigid biradical sample allowed for direct DeerAnalysis fitting without data post processing.

Contents

Glossary

$AM(t)$	Amplitude modulation function.
$B(t)$	Intermolecular background function.
B_{eff}	Effective field created by B_1 and B_0.
D	Dimensionality of the homogeneous spin distribution.
$F(t)$	Intramolecular form function.
$FM(t)$	Frequency modulation function.
$I(f)$	Inversion profile of pulse.
$K(t, r)$	Kernel function.
M_0	Magnetisation vector at equilibrium.
M_z	Longitudinal magnetisation.
N	Factor for extending 5-DEER experiment by N π 5-DEER pulses.
$P(r)$	Distance distribution.
Q_{min}	Minimal adiabaticity factor at resonance.
$S(t)$	Simulated signal.
T_1	Longitudinal relaxation time.
T_2	Transverse relaxation time.
T_{m}	Phase memory time.
$V(0)$	Echo intensity at time zero.
$V(t)$	Primary DEER trace.
Δ	Modulation depth.
$\Delta\nu$	Total sweep width of chirp pulse.
Δt	Time increment of interpulse delays.
Ω_{A}	Frequency of the DEER observer pulses.
Ω_{B}	Frequency of the DEER pump pulses.
Ω_{S}	Resonance offset.
Φ_0	Ground state electronic wavefunction.
α	Regularisation parameter.
β	Flip angle of an applied microwave pulse.
β_{HS}	Truncation factor of HS pulse.
δT	Time shift to avoid observer and pump pulse overlap.
$\delta\tau$	Fixed time shift to separate refocused and stimulated echo.
$\hat{H}_0$	Spin Hamiltonian.
$\hat{H}_{\text{DD}}$	Dipole-Dipole Hamiltonian.
$\hat{H}_{\text{EX}}$	Exchange Hamiltonian.

$\hat{H}_{EZ}$	Electron Zeeman Hamiltonian.
$\hat{H}_{HF}$	Hyperfine Hamiltonian.
$\hat{H}_{NQ}$	Nucler Quadrupole Hamiltonian.
$\hat{H}_{NZ}$	Nuclear Zeeman Hamiltonian.
$\hat{H}_{ZFS}$	Zero-Field Hamiltonian.
$\hat{\vec{I}}$	Nuclear Spin vector operator.
$\hat{\vec{S}}$	Electron spin vector operator.
$\hbar$	Reduced Planck constant, $1.054571726(47) \cdot 10^{34}$ J·s.
κ	Normalised inversion quality of pulse.
λ	Fraction of spins excited by pump pulse.
$\langle S_{Ay} \rangle$	Expectation value of the S_{Ay} operator corresponding to observable magnetisation.
μ	Macroscopic moment.
μ_0	Vacuum permeability.
$\mu_B = \beta_e$	Bohr magneton.
$\nu_1(f)$	Resonator profile.
ν_{DD}	Dipole dipole coupling in frequency units.
ν_{final}	Final frequency for chirped pulse relative to observer frequency.
$\nu_{initial}$	Starting frequency for chirp pulse relative to observer frequency.
ν_{mw}	Frequency of the microwave radiation.
ω_0	Larmor frequency.
ω_{eff}	Effective frequency.
ω_{mw}	Microwave frequency of applied pulse field.
ρ	Spin density.
σ_{echo}	Density operator at the echo position.
σ_{eq}	Density operator at thermal equilibrium.
τ	Interpulse delay.
θ	Polar angle beween $\vec{B}_0$ and $\vec{r}$.
θ, ϕ	Polar angles in PAS of g.
$\underline{1}$	Unit tensor.
$\underline{A}$	Hyperfine tensor.
$\underline{D}$	Dipolar coupling tensor.
$\underline{J}$	Exchange coupling tensor.
$\underline{g}$	g tensor.
$\vec{1}$	Unit vector.
$\vec{B}$	External static magnetic field $(0, 0, B_0)$.
$\vec{B}_0$	Static magnetic field vector.
$\vec{B}_0^T$	Transposed external magnetic field vector.
$\vec{B}_1(t)$	Pulse field.
$\vec{r}$	Spin-spin vector of length r.
ξ	Stretch parameter in relaxation law.
a_s	Pulse amplitude.

f	Instantaneous frequency.
h	Order of the HSh pulse.
k	Spin density parameter.
l	Orbital angular momentum quantum number.
m	Number of spin inversion by π pulse.
n	Order of the dynamical decoupling scheme.
p_k	Probability that pulse k inverts a spin.
q_k	$= 1 - p_k$.
t_p	Pulse length.
t_{acq}	Acquisition time.
t_{dip}	Total dipolar evolution time.
t_{max}	Maximum dipolar evolution time.
t_r	Rise time of pulse.
$\left|\frac{d\theta(t)}{dt}\right|$	Instantaneous angular velocity.
R	Distance between proton and spin density center.
$\underline{J}$	Exchange coupling tensor.
a_{iso}	Isotropic hyperfine coupling constant.
AM	Amplitude modulation.
B_0	z component of the static magnetic field vector.
CP	Carr Purcell.
CW	Continuous wave.
d	Dipole-dipole coupling.
d_8Gly	Deuterated glycerol.
d_8-OTP	Deuterated o-terphenyl.
DD	Dynamical decoupling.
DEER	Double Electron Electron Resonance.
DTT	Butan-2,3-diol-1,4-dithiol.
EPR	Electron Paramagnetic Resonance.
FID	Free induction decay.
FM	Frequency modulation.
g_e	g-value for free electron.
g_{eff}	Effective g value.
GUI	Graphical user interface.
HGly	Glycerol.
M	Macroscopic magnetisation in vector picture treatment.
MOPS	3-morpholinopropane-1-sulfonic acid.
MTSSL	(1-Oxyl-2,2,5,5-tetramethylpyrroline-3-methyl) methanethiosulfonate.
PAS	Principle Axis System.
Q	Adiabaticity factor.
S/N	Signal-to-Noise ratio.
SDSL	Site-directed spin labelling.
TWT	Traveling wave tube.

List of Figures

List of Tables

1 Introduction and Theory

1.1 Motivation of research

The pulsed Electron Paramagnetic Resonance (EPR) Double Electron Electron Resonance (DEER) experiment allows to determine distance distributions within electron spin pairs introduced in biomolecules by site-directed spin labelling (SDSL) in the nanometer range. The obtainable distance range and sensitivity of the DEER experiment strongly depends on the transverse relaxation of the electron spin as this determines the maximal available dipolar evolution time in the DEER experiment. DEER experiments on nitroxides are typically carried out in the low temperature (50 K) and low concentration regime (< 50 μM) where relaxation is mainly induced from fluctuating hyperfine fields.[1] These fluctuations result from nuclear spin diffusion and lead to a decay function $\exp\left[-\left(\frac{t}{T_m}\right)^\xi\right]$ with phase memory time T_m and stretch parameter $\xi > 1$.[2] More specificially, in non-deuterated samples protons present in the solvent, lipid or the protein itself give rise to proton spin diffusion.[1] The phase relaxation can be prolonged by deuteration, an approach which is however not always feasible for a specific biological sample. Thus, proton spin diffusion is especially deterimental for the conformational study of membrane proteins by DEER.

Electron spin coherence losses due to the above described spin diffusion processes can be reduced by multiple refocusing as $\exp\left[-\left(\frac{t/n}{T_m}\right)^\xi\right]^n < \exp\left[-\left(\frac{t}{T_m}\right)^\xi\right]$ for n refocusing pulses, which is the basis of dynamical decoupling (DD) schemes.[3]

In this Master thesis, the relaxation behaviour of the nitroxide spin echo created by different DEER observer sequences is studied for model systems with spin environments typical for DEER applications. The relaxation pulse sequences were derived from

DD theory including Carr-Purcell (CP) and Uhrig schemes to assess the feasibility of applying corresponding n^{th} order DD-DEER experiments for the particular spin environment.

Higher-order DD-DEER experiments, first introduced by Borbat, Georgieva and Freed[4], require multiple pump pulses relative to the standard 4-DEER scheme of order $n = 1$. In these experiments, complications arise due to incomplete excitation of pump spin packets.[5]

For minimisation of unwanted signal contributions different adiabatic inversion pump shapes are tested both at X- and Q-band frequencies after assessing their inversion performance by simulation. Finally a combination of DD-DEER observer sequences with efficient pump inversion pulses was used to extend the dipolar evolution time with decreased signal artifacts relative to a rectangular pump pulse DD-DEER scheme. This implies a possible extension of measureable interspin distances in biomolecules which were otherwise out of reach due to spin echo relaxation.

1.2 Spin Hamiltonian

The spin Hamiltonian $\hat{H}_0$ contains all interactions within a paramagnetic system between electron spins, nuclear spins and the magnetic field in the presence of an outer magnetic field.[6]

$$\hat{H}_0 = \hat{H}_{EZ} + \hat{H}_{HF} + \hat{H}_{NZ} + \hat{H}_{NQ} + \hat{H}_{ZFS} + \hat{H}_{EX} + \hat{H}_{DD} \tag{1}$$

with $\hat{H}_{EZ}$ and $\hat{H}_{NZ}$ being the electron and nuclear Zeeman interactions, which describe the interaction between the static magnetic field and the electron or nuclear spin, respectively. $\hat{H}_{HF}$ represents the hyperfine couplings between electron and nuclear spins. For nuclei with $I > 1/2$ the nuclear quadropole interaction $\hat{H}_{NQ}$ corresponds to the nuclear spin interacting with an electric field gradient. The electron-electron spin interaction can be separated into a strong contribution for $S > 1/2$, called the zero-field term $\hat{H}_{ZFS}$ and two types of weaker interaction. The latter include the exchange interaction $\hat{H}_{EX}$ and the dipole-dipole interaction $\hat{H}_{DD}$.[6]

In the following only interactions relevant for this thesis are discussed in more detail.

1.2.1 Electron Zeeman interaction $\hat{H}_{EZ}$

The electron Zeeman term describes the interaction between the electron spin and the external magnetic field B_0 for which the Hamiltonian $\hat{H}_{EZ}$ reads[6]

$$\hat{H}_{EZ} = \frac{\mu_B}{\hbar} \vec{B}_0^T g \hat{\vec{S}} \tag{2}$$

with the Bohr magneton $\mu_B = \beta_e = 9.27400899(37) \cdot 10^{-24}$ JT^{-1}, reduced Planck constant $\hbar = 1.054571726(47) \cdot 10^{-34}$ Js , the transpose of the external magnetic field vector $\vec{B}_0^T$, the g tensor g and the electron spin vector operator $\hat{\vec{S}}$. The external magnetic field vector can be expressed in the principle axis system of the g tensor[6]

$$\vec{B}_0 = B_0 \left[\sin(\theta)\cos(\phi), \sin(\theta)\sin(\phi), \cos(\theta) \right] \tag{3}$$

with polar angles θ and ϕ.

In case of an anisotropic g tensor this gives rise to the electron Zeeman interaction being dependent on the molecular orientation relative to the external magnetic field. Therefore an effective g value can be computed[6]

$$g_{eff} = \sqrt{g_x^2 \sin^2(\theta)\cos^2(\phi) + g_y^2 \sin^2(\theta)\sin^2(\phi) + g_z^2 \cos^2(\theta)} \tag{4}$$

with the principle g values g_x, g_y and g_z. The high-field approximation assumes that the electron Zeeman interaction dominates the spin Hamiltonian. Thus, the corresponding resonance field $B_{0, res}$ at each molecular orientation is specified by θ and ϕ[6]

$$B_{0, res} = \frac{h\nu_{mw}}{g_{eff}\mu_B} \tag{5}$$

Equation (5) implies that for a constant microwave frequency ν_{mw} at a certain field B_0 a subset of molecular orientations is excited which is characterised by g_{eff}.

1.2.2 Hyperfine interaction $\hat{H}_{HF}$

The hyperfine interaction term can be written as[6]

$$\hat{H}_{HF} = \sum_k \hat{\vec{S}}^T \underline{A}_k \hat{\vec{I}}_k \tag{6}$$

applicable to all k nuclei with $I_k > 0$. $\hat{\vec{I}}_k$ corresponds to the nuclear spin vector operator for the above specified nuclei in combination with the hyperfine tensors $\underline{A}_k$. The total hyperfine tensor for each nucleus k can be obtained from[6]

$$\underline{A}_k = a_{iso,k} \underline{1} + \frac{g \underline{T}_k}{g_e} \tag{7}$$

with the unit tensor $\underline{1}$, the isotropic coupling a_{iso} and the hyperfine tensor $\underline{T}_k$.

The hyperfine interaction arises from an anisotropic through-space dipolar coupling between the magnetic moments of the electron and nuclear spin but can also be caused by the isotropic Fermi contact interaction. The latter results from the probability of locating an electron in an s orbital at the nucleus position. This leads to an expression for the isotropic coupling $a_{iso} \hat{\vec{S}}^T \hat{\vec{I}}$ being dependent on the squared ground state electronic wave function at the position of the nucleus $|\Phi_0(0)|^2$ for the s orbital.[6]

Electrons in orbitals characterised by orbital angular momentum quantum number $l > 0$ lead to anisotropic dipole-dipole interactions which can be specified in a hyperfine coupling tensor $\underline{T}$. Protons are the dominating hyperfine coupling partner for the electron spin in biomolecules for which the anisotropic contribution to the A_k term reduces to a point-dipole approximation as protons spin densities in orbitals with $l > 0$ can be neglected.[2] Hence, the hyperfine tensor $\underline{T}_k$ simplifies to a sum over all centers with spin densities ρ_j with which the proton s orbitals couple at a distance R_j between the proton and the spin density center [6]

$$\underline{T}_k = \frac{\mu_0}{4\pi\hbar} g_e \mu_B g_n \mu_n \sum_{j \neq k} \rho_j \frac{3\vec{n}_j \vec{n}_j^T - \vec{1}}{R_j^3} \tag{8}$$

with the unit vector $\vec{1}$ and $\vec{n}_j$ unit vectors along the proton to the spin density center.

1.2.3 Electron-electron spin interactions $\hat{H}_{EX}$ and $\hat{H}_{DD}$

The electron-electron spin interactions described by $\hat{H}_{EX}$ and $\hat{H}_{DD}$ are treated differently depending on the coupling strength. In the high-field approximation, the electron Zeeman term dominates the spin Hamiltonian interactions. If the exchange coupling as well as the dipole-dipole coupling are smaller than the difference between the Zeeman frequencies, the two spins are considered to couple weakly. In this regime the system can be described based on the individual spins and the secular components of the exchange and dipole-dipole interaction suffice. [3]

1.2.3.1 Exchange interaction $\hat{H}_{EX}$

The exchange interaction term reads [6]

$$\hat{H}_{EX} = \hat{\vec{S}}_A^T \underline{J} \hat{\vec{S}}_B \tag{9}$$

with the exchange coupling tensor $\underline{J}$. Exchange contribution to the spin Hamiltonian is relevent, when orbitals of the two spins overlap significantly, so that electrons exchange. As the extension of the orbitals into space decays exponentially, this term can be neglected as dipole-dipole coupling dominates at distances longer than 1.5 nm, provided there is no extensive spin delocalisation. There exists no classical analogue to exchange interaction.

5

1.2.3.2 Dipole-dipole interaction $\hat{H}_{DD}$

The dipole-dipole Hamiltonian $\hat{H}_{DD}$ can be derived from the classical description for the interaction energy of two magnetic dipoles via the correspondence principle by substituting the magnetic moments by related spin operator times $\hbar\gamma$.

$$\hat{H}_{DD} = \hat{S}_A^T \underline{D} \hat{S}_B = \frac{1}{r^3} \cdot \frac{\mu_0}{4\pi\hbar} \cdot g_A g_B \mu_B^2 \left[\hat{S}_A \hat{S}_B - \frac{3}{r^2} \left(\hat{S}_A \vec{r} \right) \left(\hat{S}_B \vec{r} \right) \right] \tag{10}$$

with the dipole-dipole coupling tensor $\underline{D}$, the vacuum permeability μ_0 and the spin-spin vector $\vec{r}$ of length r.[6]

In the high-field approximation the magnetic moment of the two nitroxide spins can be assumed to be parallely aligned along the external magnetic field. This in turn allows to characterise the molecular orientation with respect to the external magnetic field by a single polar angle θ between $\vec{B}_0$ and $\vec{r}$.

For the weakly coupled electron spins the dipolar interaction simplifies to its secular part with the dipole-dipole coupling d[6]

$$\hat{H}_{DD} = d\hat{S}_{Az}\hat{S}_{Bz} = \omega_{DD}(1 - 3\cos^2\theta)\hat{S}_{Az}\hat{S}_{Bz} = \frac{\mu_0}{4\pi} \frac{g_A g_B}{r^3} \frac{\mu_B^2}{\hbar} \hat{S}_{Az}\hat{S}_{Bz} \tag{11}$$

For organic radicals, i.e. nitroxide spin labels, g_A and g_B can be approximated by the g-value for the free electron $g_e = 2.00231930$ such that the dipole-dipole coupling expressed in frequency units simplifies to[6]

$$\nu_{DD} = \frac{\omega_{DD}}{2\pi} = \frac{52.04\text{MHz}}{r^3 \cdot \text{nm}^{-3}} \tag{12}$$

1.3 Pulses in EPR experiments

A theoretical treatment of EPR requires a quantum mechanical model, as quantum objects are treated on a microscopic level. However, as pulsed EPR experiments involve an ensemble of spins, the magnetisation vector picture can often be employed. In this picture, the effect of nanosecond microwave pulses on a spin ensemble can be described.[7] Pulsed EPR techniques allow for the separation of interactions where the macroscopic magnetisation M is the observable.

1.3.1 Magnetisation vector picture

M can be obtained from the vector sum over all magnetic moments μ of the electrons present in the sample volume. When the spin system is equilibrated the magnetisation vector M_0 is aligned with the external static magnetic field $\vec{B} = (0, 0, B_0)$. If M is not perfectly aligned with $\vec{B}$ a torque perpendicular to M leads to the precession of the magnetisation vector around the z-axis with the so called Larmor frequency ω_0 following the classical equation of motion[7]

$$\frac{d}{dt}\vec{M} = \vec{M} \times \frac{-g_e\beta}{\hbar}\vec{B} \tag{13}$$

To deflect the magnetisation vector from its equilibrium position $\vec{M}_0$ a linearly polarised magnetic field $B_1(t)$ perpendicular to the external magnetic field vector $\vec{B}$ is applied which is characterised by a frequency ω_{mw}. The $\vec{B}_1(t)$ field is conveniently treated in a rotating frame to remove its time dependency. This alters the Larmorfrequency to a so called resonance offset Ω_S being defined as

$$\Omega_S = \omega_0 - \omega_{mw} \tag{14}$$

Applying a microwave field of magnitude B_1 creates an effective field B_{eff} together with the static field $\vec{B}$ along the z direction. The magnetisation vector precesses around B_{eff} with a frequency ω_{eff}[7]

$$\omega_{eff} = \sqrt{\Omega_S^2 + \omega_1^2} \tag{15}$$

a phenomenon called nutation. Where ω_1 relates to B_1 by[7]

$$\omega_1 = \frac{g_e \beta_e B_1}{\hbar} \tag{16}$$

An angle θ between the effective field and the z axis can be defined by[7]

$$\theta = \arctan\left(\frac{\omega_1}{\Omega_S}\right) \tag{17}$$

For an on-resonant pulse ($\omega_{mw} = \omega_0$) the nutation frequency ω_{eff} corresponds to ω_1, such that $\theta = 90°$. A pulse of length t_p will rotate the magnetisation vector through the angle β which is related to ω_1 by[7]

$$\beta = \omega_1 \cdot t_p \tag{18}$$

Depending on the choice of t_p, $\vec{M}_0$ can for example be inverted by a π pulse whereas a $\pi/2$ pulse along the x direction flips the initial magnetisation vector $\vec{M}_0$ into the xy plane as shown for the first pulse in Fig. 1a.

During the time delay τ, the magnetisation vector precesses with ω_1 which can be described in terms of the magnetisation components: M_z corresponds to the longitudinal magnetisation, while M_x and M_y are the transverse magnetisation. The oscillatory signal of the latter is detected by a phase-sensitive detector and called free induction decay (FID). In this vector picture a pulse sequence applied to a spin ensemble can be envisioned as a combination of rotations and precessions of the magnetisation vector as illustrated for the primary echo pulse scheme in Fig. 1a on page 12.[7]

1.3.2 Adiabatic pulses

For $\omega_{mw} \neq \omega_0$ a pulse is called off-resonant, and the nutation frequency corresponds to ω_{eff}. Where $\Omega_S \ll \omega_1$ implies $\theta \approx 0$. Hence, pulses can only achieve an effective inversion over a desired bandwidth if the condition $\omega_1 \gg \Omega_S$ is satisfied. This condition is limited by availability of field strengths.[8]

A monochromatic rectangular pulse is only amplitude modulated with sharp edges at the pulse extremities. Inversion efficiency over a broad bandwidth can be improved by adiabatic pulses which are characterised by a modulation of their frequency (FM) and amplitude (AM).

The working principle of an adiabatic pulse is based on dragging the magnetisation vector along with the effective field ω_{eff} upon application of the frequency modulation. This can only be achieved if the adiabatic theorem is fulfilled which can be defined in terms of an adiabaticity factor $Q^{[9]}$, given by

$$Q(t) = \frac{\omega_{\text{eff}}}{\left|\frac{d\theta(t)}{dt}\right|} \gg 1 \tag{19}$$

with the instantaneous angular velocity $\left|\frac{d\theta(t)}{dt}\right|$ of the nutation frequency ω_{eff}.

During the frequency sweep, the condition $\omega_{\text{eff}} = \omega_0$ will be satisfied where the pulse becomes on resonant and the adiabaticity factor reaches a minimum Q_{min} [9]

$$Q_{\text{min}} = \frac{2\pi \nu_1^2 t_{\text{p}}}{\Delta \nu} \tag{20}$$

with $\nu_1 = \frac{\omega_1}{2\pi}$ and the total sweep width $\Delta\nu$ both in Hz.

The class of adiabatic pulses comprises different pulse shapes being specified by their respective $FM(t)$ and $AM(t)$ functions. Generally, the adiabaticity factor Q can be expressed in terms of the AM and FM functions. [10]

From this relation the $FM(t)$ function for each amplitude modulation function $AM(t)$ can be derived by the so-called offset-independent adiabaticity. It has been shown that an adiabaticity value of $Q(t) = 5$ leads to efficient inversion. [9] The offset-independent adiabaticity principle requires that for each instantaneous frequency f within the pulse frequency range $\Delta\nu$ the same $Q(t) = Q_{\text{min}}$ (i.e. 5) applies [8], so that

$$FM(t) = \frac{1}{Q_{\text{min}}} \int_0^t AM^2 d\tau \tag{21}$$

In the following pulse shapes employed for the pump sequence of the DEER experiments are discussed, where $FM(t)$ can be derived from the stated $AM(t)$ by using equation (21).

1.3.2.1 Linear-chirp pulse

The linear-chirp pulse can be described by a rectangular pulse $AM(t)$ function on which a quarter sine apodisation window is applied. The latter is characterised by its rise time t_r.

The pulse can be further adjusted by choice of t_p and its frequency range $\Delta\nu$ starting from $\nu_{initial}$ to ν_{final} resulting in an adiabaticy factor as specified in 20. [9]

1.3.2.2 Hyperbolic secant pulse

The amplitude of the hyperbolic secant pulse (HS) is defined by [10]

$$AM_{HS}(t) = \text{sech}\left(\frac{t\beta_{HS}}{t_p}\right) \tag{22}$$

where the parameter β_{HS} allows for truncation of the otherwise non-zero amplitude of the hyperbolic secant waveform for $t \neq \infty$ at a chosen minimum amplitude.

The HS pulse gives an adiabaticy factor of [11]

$$Q_{min} = \frac{4\pi\nu_1^2 t_p}{\beta_{HS}\Delta\nu} \tag{23}$$

1.3.2.3 Hyperbolic secant pulse of order h

Hyperbolic secant pulse of order h (HSh) are characterised by an amplitude modulation function

$$AM_{HSh}(t) = \text{sech}\left[\left(\frac{t\beta_{HS}^x}{t_p}\right)^h\right] \tag{24}$$

The HSh pulse becomes increasingly broader with respect to a HS pulse for higher values of order h. The parameter x allows to compensate for the effect of order h on β_{HS}. The value for Q_{min} has to be evaluated numerically for each order of h as a function of AM and FM.

1.3.3 Inversion Characterisation

A π pulse aims to flip the magnetisation vector by 180°. For the initial magnetisation vector $\vec{M}_0 = (0, 0, +1)$ a complete 180° inversion leads to $\vec{M} = -\vec{M}_0 = (0, 0, -1)$. Hence the inversion I of the magnetisation vector can be characterised via its z component M_z

$$I = 1 - \frac{M_z + 1}{2} \tag{25}$$

This allows to describe a pulse by its inversion profile $I(f)$ with $I(f) = 0$ corresponding to no inversion and $I(f) = 1$ to complete inversion at each instantaneous frequency f. For a pump pulse as employed in DEER sequences, the quality of the pump pulse shape can be characterised by the normalised inversion quality κ

$$\kappa = \frac{4 \int I(f) \, (1 - I(f)) \, df}{\int df} \tag{26}$$

with integration limits covering the frequency range for which $I(f)$ deviates significantly from zero. Considering equation (26) the κ value decreases with better inversion quality down to $\kappa = 0$ which corresponds to complete inversion.

1.4 Pulsed EPR experiments

1.4.1 Primary and stimulated echo

Short effective relaxation time of the FID leads to signal decay within the deadtime after the microwave pulse. Hence EPR favors echo detection over FID detection. Echo detection works by creating a refocused transverse magnetisation which is called an electron spin echo as depicted in Fig. 1a.[7]

The electron spin echo is created by a primary echo sequence where the dephasing of the transverse magnetisation after the first $\pi/2$ pulse is represented by the vector a and vector b in Fig. 1a. The two arrows correspond to spin packets with positive and negative resonance offset due to interaction anisotropy. Applying a π pulse placed between two interpulse delays of equal lengths refocuses the transverse magnetisation at time 2τ.

A stimulated echo experiment with the same total length of 2τ can be derived from the primary echo sequence by splitting the refocusing π pulse into two $\pi/2$ pulses applied after $\tau/2$ and $3\tau/2$ respectively and depicted in Fig. 1b. The resulting stimulated echo at 2τ relaxes slower in comparison to the refocused echo and has a different phase. This relaxation behaviour is related to the creation of longitudinal magnetisation M_z by the second $\pi/2$ in the stimulated echo sequence. While M_z decays with T_1, transverse magnetisation relaxes with T_2 while $T_2 < T_1$ generally applies.[7]

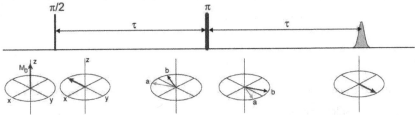

(a) Primary echo sequence with magnetisation vector illustration. The first $\pi/2$ pulse rotates equilibrium magnetisation M into the xy plane where it dephases. The π pulse refocuses the magnetisation at time 2τ. This is also called a Hahn echo.

(b) Stimulated echo sequence.

Fig. 1: Primary and stimulated echo sequences.

1.4.2 Nutation experiment

The nutation experiment as shown in Fig. 2 allows to determine the nutation frequency ν_1 for an applied mw pulse of a certain frequency f. From this the flip angle β achieved by the pulse can be obtained.

In the experiment the pulse length t_1 of the first nutation pulse is varied and the echo intensity is detected after $t_1 + T + t_{\pi/2} + t_\pi + 2\tau$ as a function of t_1. The point in time at which the first minimum of the detected echo intensity is reached allows to determine the pulse length t_p for a π pulse. [7]

Fig. 2: Pulse sequence to determine the nutation frequency by detecting the echo intensity as a function of pulse length t_1 of the intial pulse.

1.4.3 DEER sequences

The pulsed DEER experiment allows to isolate the dipolar interaction information between two electron spins which is otherwise hidden within the natural linewidth of an EPR signal. [6] Equation (12) relates the distance r between two spins to the strength of their dipolar interaction in frequency units.

1.4.3.1 Computation of echo amplitude modulation function

Under the high-field approximation and in the weak electron spin dipolar coupling limit the Hamiltonian relevant for all DEER experiments reads

$$\hat{H}_0 = \Omega_A \hat{S}_{Az} + \Omega_B \hat{S}_{Bz} + \omega_{DD} \left(1 - 3\cos^2\theta\right) \hat{S}_{Az}\hat{S}_{Bz} \qquad (27)$$

If the observer DEER sequence refocuses the resonance offset Ω_A the respective DEER scheme can be conveniently treated in a rotating frame such that the electron Zeeman term of the A spin in expression (27) can be disregarded. Hence, product operator formalism can be used to propagate the spin system under the secular part of the dipole-dipole interaction for the free evolution periods.

Therefore the density operator at the echo position σ_{echo} can be computed starting from an initial density operator for the observer spin at thermal equilibrium $\sigma_{eq} = -\hat{S}_{Az}$. From σ_{echo} the modulation of the observable magnetisation $\langle S_{Ay} \rangle$ can be obtained.

Alternatively, the signal oscillation $\frac{V(t)}{V(0)}$ can be derived more intuitively by simple inspection of the pulse sequence. As each π pulse inverts the sign of the dipolar evolution for either the observer or pump spins, the total dipolar evolution time t_{dip} can be easily computed from alternating subtraction and addition of sequential interpulse delays. Starting with an initial positive delay after the first $\pi/2$ pulse leads to a signal oscillation with $\cos\left[\frac{d \cdot t_{dip}}{2}\right]$.

1.4.3.2 4-pulse DEER sequence

Since the introduction of the 4-DEER experiment, the DEER-signal can be obtained free of dead time.[12] The experiment relies on two mw frequencies which excite two coupled electron spins separately. Interspin distances between two spins with identical spectra can be obtained by choosing excitation bands of the mw pulses so that distinct populations of the spectra are excited.

The observer sequence at frequency ω_A of the DEER experiment places a refocused primary echo at $t_{\pi/2} + t_\pi + 2t_1$ which corresponds to time zero of the dipolar evolution time t_{dip}. At $t_{\pi/2} + 2t_\pi + 2t_2$ the primary echo is refocused as shown in Fig. 3a. At the pump frequency ω_B the π pump pulse acts on the B spins with an incremented pump pulse delay t while the observer sequence remains static. This feature furthermore allows the exclusion of spin echo decay from the dipolar time evolution.[13]

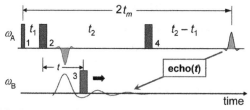

(a) 4-DEER with t_2 long dipolar modulated trace (——).

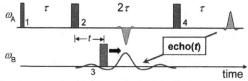

(b) 4-DEER sequence with CP-type observer sequence ($t_2 = 2t_1 = 2\tau$). The DEER trace (——) is shifted with respect to the $t_2 \neq 2t_1$ setup. Yet only half of the 2τ long dipolar signal can be used as the other half contains redundant information due to the symmetric nature of the trace.

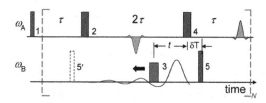

(c) 5-DEER sequence which allows to use the 2τ long dipolar modulated trace which is time reversed relative to Fig. 3a.

Fig. 3: Comparison of the 4-DEER and 5-DEER sequence. The figure is taken from a paper by Borbat, Georgieva and Freed.[4]

The application of the π pump pulse inverts the state of the B spin, which alters the local field created by spin B at the site of spin A. This leads to a change in spin A frequency by the dipolar coupling ω_{DD}. As the π pump pulse delay t is incremented the refocused echo signal at $t_{\pi/2} + 2t_\pi + 2t_2$ is modulated with ω_{DD} as a function of time.

As only a fraction $\lambda_i < 1$ of all spins B is excited by the pump pulse, a fraction λ_i of the spin A magnetisation experiences a phase gain. The echo amplitude as a function of the pump pulse increment t can be expressed as the product of all coupled spins B_i to A and reads[14]

$$V(t) = \prod_i \left\{ 1 - \lambda_i \left[1 - \cos\left(\omega_{\text{DD},i} t\right) \right] \right\} \tag{28}$$

where t runs from 0 to the maximum dipolar evolution time t_{max}. For the standard DEER experiment $t_{\text{max}} = t_2$ as illustrated in Fig. 3a. The echo intensity is modulated by $\cos\left(\omega_{\text{DD},i} t\right)$ which can be derived as described in section 1.4.3.1.

Assuming a isolated spin pair, such that one observer spin A is coupled only to one pump spin B within the reachable DEER distance and considering all other spins B_i homogeneously distributed in space, the effective signal can be obtained by integrating over all possible spin vector orientations[14]

$$V(t) = \left\{ 1 - \lambda \left[1 - \int_0^1 \cos\left(\omega_{\text{DD}} \left(1 - 3\cos^2(\theta)\right) t \right) d\cos(\theta) \right] \right\} B(t) = F(t)B(t) \tag{29}$$

with intermolecular background function $B(t)$ and intramolecular form factor $F(t)$. The former arises from cases where the semi-isolated spin pair assumption breaks down and additional spins B are found within the DEER accessible distance from observer spin A. The Fourier transformation of the form factor yields a Pake pattern.[6]

The modulation depth Δ of the DEER trace is defined by the amplitude difference of $F(t)$ between $t = 0$ and a time t where the dipolar modulation has decayed completely. This means that $F(t)$ decays to $1 - \Delta \neq 0$ which arises from incomplete excitation of the pumped spins. Assuming that no geometrical correlation between the electron spins in the pair exists and that not more than two spins are deterministically coupled, Δ is determined by λ. Similiarly, the DEER signal intensity at $t = 0$ corresponds to the fraction of spins excited by the observer pulses.[15]

16

The maximal obtainable distance is limited by the transverse relaxation time T_2, or, more precisely, phase memory time T_m of the electron spin.

Typical T_2 values in the microsecond regime allow the measurement of dipole-dipole couplings of 100 kHz which translates into an upper limit of 8 nm for DEER accessible distances. This upper distance limit is lowered for soluble proteins and membrane proteins, in particular, as their natural spin environment leads to increased spin echo relaxation which implies a limitation in t_{max}.[6]

1.4.3.3 5-pulse DEER sequence

The 5-DEER sequence as published by Borbat, Georgieva and Freed[4] allows to extend the maximum dipolar evolution time t_{max} relative to the 4-DEER experiment. Fig. 3 illustrates the working principle: Firstly, the pulse sequence at the observer frequency ω_A is symmetrised to a CP-type ($\tau - \pi - 2\tau - \pi$)-scheme in order to minimise transverse relaxation for echo decaying with a stretch parameter $\xi > 1$ as later defined in equation (33). This allows for an extension of t_{max} to 2τ.

The echo modulation function for the 4-DEER as specified in equation (28) is symmetric with respect to $t = t_1$. For the CP-type 4-DEER setup as shown in Fig. 3b $t_2 = 2t_1$ applies which makes half of the 2τ long DEER trace redundant as it contains the identical distance information. Upon addition of a second fixed pump pulse (5) the dipolar modulated trace is shifted to represent an extended but time reversed trace relative to the standard 4-DEER signal illustrated in Fig. 3c. Placing the fixed pump pulse at the alternative (5') position and pushing pump pulse (3) from left to right allows to record the dipolar trace more conveniently as the signal zero time appears earlier.

The addition of a second pump pulse introduces complications as the second pump pulse flips spins only with probability p_5 at ω_B, whereas a pump pulse k does not lead to an inversion with q_k.

Hence, an expression for the intramolecular part of the 5-DEER echo amplitude modulation can be derived[4]

$$V^{5\text{-DEER}}(t) \propto \langle q_3 q_5 + q_3 p_5 \cos\left(d\left(\tau - \delta T\right)\right) + p_3 q_5 \cos\left(d\left(\tau - t\right)\right) + p_3 p_5 \cos\left(d\left(t - \delta T\right)\right) \rangle \quad (30)$$

where only a small pulse excitation overlap at ω_A and ω_B was assumed and the dipolar coupling during the pulses was neglected. The angular brackets in expression (30) imply averaging over all p_k, q_k and all orientations hidden in d.

In expression (30) only two terms depend on the pump pulse delay t where the term $p_3 p_5 \cos\left(d\left(t - \delta T\right)\right)$ corresponds to the desired 5-DEER signal and the third term with $p_3 q_5 = p_3(1 - p_5)$ corresponds to an unwanted residual 4-DEER signal due to the effectively absent second pump pulse. The basic 5-DEER sequence as shown in Fig. 3c can be further extended by replicating the scheme within the pink brackets. Keeping the total sequence length constant, a setup with an observer sequence $\pi/2 - (\tau/n - \pi - \tau/n)$-echo corresponds to a 5-DEER sequence for $n = 2$. The order n of the DD-DEER experiment allows to define a 7- and 9-pulse DEER setup corresponding to order $n = 3$ and $n = 4$ respectively.

1.4.3.4 Pump pulses in DEER

The relative weight of the residual 4-DEER signal ($p_3 q_5$) with respect to the desired 5-DEER signal ($p_3 p_5$) in expression (30) is therefore related to the inversion efficiency of a pump pulse. Hence higher order DD-DEER sequences require close approximations to rectangular excitation profiles for the pump pulses such that spin packages within the pump band are inverted whereas leaving the spins outside this region unaffected. The κ parameter defined in equation (26) allows to quanitify the inversion quality of a pump pulse.

Spindler and Prisner suggested to increase inversion efficiency by applying adiabatic inversion pulses instead of monochromatic rectangular pump pulses.[5] The latter are non-selective as their excitation profile in the frequency domain corresponds roughly to a sinc function which implies excitation beyond their formal bandwidth. Furthermore, sharp edges at the extremities of the pulse lead to transient phase modulation during the rising and falling time due to hardware non-linearity and limited bandwidth.[16] In contrast, adiabatic pulses have been shown to increase the inversion bandwidth leading to increased modulation depth of the DEER signal. Moreover, the approximation of the ideal rectangular excitation profile by adiabatic inversion pulses allow to minimise the overlap of pump and observer excitation bands without the need to sacrifice sensitivity or dipolar evolution time. The last two approaches are otherwise commonly employed[14] to suppress overlap artifacts which arise from 2+1 train experiment spin dynamics of the observers spins.[17] Additionally to these signal artifacts typically arising at the end of the DEER trace, band overlap may also lead to nuclear modulation which originates from excitation of forbidden electron-nuclear transitions.[12] Typically the nuclear Zeeman frequencies of surrounding protons (14 MHz) and deuterons (2 MHz) are observed at X band. At Q-band frequencies proton modulations are less noticable while a nuclear Zeeman frequency of 8 MHz indicates the presence of matrix deuterons in the sample.[14]

Besides the advantages of adiabatic shaped pulses presented here, the nature of these pump pulses introduces an inversion time dispersion. As the frequency is swept the inversion time for each spin packet varies as a function of its respective resonance offset Ω_B. Only for an even number of pump pulses corresponding to the order n of the DD DEER experiment, the dispersion can be refocused by employing pump pulse pairs of opposite frequency sweep directions.[5] Therefore only DD-DEER experiments of even order n are considered in the following which excludes the 7-DEER scheme corresponding to order $n = 3$.

1.4.3.5 DEER signal

The evolution of a spin system under the Hamiltonian specified in equation (27) for any DEER sequence allows to obtain the modulation function for the echo amplitude as a function of total pump pulse delay t. The damping of the echo intensity seen experimentally occurs due to the distribution of angle θ and distance r which translates into a variation of ω_{DD}. The destructive interference of the different frequencies leads to a damping of the echo modulation. Secondly, the magnetisation propagates in the xy plane such that the final echo amplitude is reduced due to transverse relaxation for pulse sequence duration.

1.4.4 DEER data analysis

The primary 4-DEER trace $V(t)$ contain intra- and intermolecular signal contributions referred to as form factor $F(t)$ and background function $B(t)$, defined in equation (29). Extraction of the intramolecular distance distribution is only possible upon reliable separation of $F(t)$ from $V(t)$ with knowledge of the functional form describing the background contribution. In the most simple case this is assumed to be expressed by

$$B(t) = \exp\left(-kt^{D/3}\right) \tag{31}$$

with spin density parameter k and the dimensionality of the homogeneous spin distribution D.[18]

Only if intramolecular distances are much shorter than the intermolecular distances, the two signal contributions dominate the primary data in different parts of the dipolar time and hence can be distinguished. Therefore t_{max} of the DEER trace has to extend beyond the time at which the dipolar modulation has decayed. In the absense of orientational selection, there exists a unique mapping of the orientationally averaged $F(t)$ and the distance domain described by a distance distribution $P(r)$. The precision with which

$P(r)$ can be obtained depends on t_{max} of the DEER trace. It has been shown that the extraction of the mean distance as well as the standard deviation of $P(r)$ requires at least two periods of the dipolar evolution function.[19]

Due to the ill-posed nature of extracting a distance distribution $P(r)$ from the form factor $F(t)$, Tikhonov regularisation needs to be employed. A simulated signal $S(t)$ based on[18]

$$S(t) = K(t,r)P(r) \tag{32}$$

with the Kernel function $K(t,r)$ is expected to fit the form factor. $K(t,r)$ is analytically known for ideal 4-DEER experiments. The regularisation approach is based on a compromise between smoothness of $P(r)$ and mean square deviation between the $S(t)$ from $F(t)$. The regularisation parameter α determines the relative weight of the two criterions. Small α values are required for narrow distance distributions whereas larger α values lead to broader peaks in $P(r)$ but also to better stabilisation against noise.[18]

1.5   Relaxation

After the excitation by a microwave pulse the longitudinal magnetisation relaxes exponentially to its equilibrium magnetisation M_0 with the so called spin-lattice relaxation time T_1. Similiarly, the transverse relaxation time T_2 characterises the decay of the M_x and M_y components. Therefore T_1 and T_2 describe the relaxation of a simple two-level spin system in the absence of a radiation field.

For frozen glassy solution the phase memory time T_m is a more appropriate parameter to characterise the loss of electron spin phase coherence because the system treated corresponds to a network of coupled spins instead of a simple two-level system for which T_2 is defined.

21

1.5.1 Relaxation processes in immobilised samples

The underlying mechanism for the transverse magnetisation decay determines the shape of the echo decay curve where the echo intensity as a function of time t can be generally fitted using a stretched exponential function

$$V(t) = V(0) \cdot \exp\left[-\left(\frac{t}{T_{\mathrm{m}}}\right)^{\xi}\right] \tag{33}$$

$V(0)$ being the echo intensity at time zero and ξ the stretch exponent[2]. Processes which may contribute to the relaxation of the transverse magnetisation include[20]

- Molecular motion of the spin label

- Dipolar interaction between electron spins

  - Instantaneous Diffusion (ID): At sufficiently high spin concentration inversion of coupled spins in the observer's proximity leads to a stochastic change of the local field at the observed spins.

  - Spectral Diffusion (SD): Incorporates the coupling of pulse-excited and non-excited spins due to either energy exchange with the lattice or flip-flop exchange processes (spin diffusion).

- Dipolar interaction between electron spins with surrounding nuclei

  - Hyperfine Interaction (HF)

For $\xi = 1$ in equation (33) the decay is simple exponential and $T_{\mathrm{m}} = T_2$. In the low temperature and low spin concentration regime it has been shown that the relaxation mechanism of nitroxide spins is dominated by fluctuating hyperfine fields which originate from nuclear spin diffusion of the matrix protons.[1] The actual mechanism involves a change in the hyperfine field of the protons due to electron spin excitation. The resulting proton magnetisation is no longer in equilibrium and diffuses via dipole-dipole coupling. More specifically, the flip-flop transition of the dipole-dipole coupling is responsible for this proton spin diffusion such that the hyperfine field fluctuates due to flipping protons. Deuterons have a smaller magnetic moment in comparsion to protons such that deuteration reduces the contribution of fluctuating hyperfine fields to the transverse magnetisation relaxation.

Hence matrix deuteration is a popular method to increase T_m in biomolecular samples. However this procedure is limited as the change in matrix composition may alter the structure which is particularly the case for biomacromolecules. Moreover, procedures to deuterate the objects themselves may be either impossible or too costly.[21]

1.6 Dynamical Decoupling

Dynamical decoupling schemes aim to suppress decoherence. The traditional building block is characterised by a $(\tau - \pi - \tau)$ sequence which refocuses static but non-uniform couplings in a spin system.

For systems with $\xi > 1$ spin diffusion enhances the natural T_2 relaxation. This effect can be alleviated by n fold refocusing in between adjacent interpulse delays of length τ/n. This dynamical decoupling scheme was introduced by Carr and Purcell (CP).[3] It decreases the relative contribution of diffusion-induced relaxation for a constant sequence length of 2τ relative to the T_2 dependent relaxation for increasing number of n. More formally this can be expressed as

$$\exp\left[-\left(\frac{t/n}{T_m}\right)^{\xi}\right]^{n} < \exp\left[-\left(\frac{t}{T_m}\right)^{\xi}\right] \tag{34}$$

Traditionally, dynamical decoupling schemes have been based on equidistant π pulses as in CP schemes. More recently, Uhrig has suggested an optimised decoupling scheme in which the π pulse delays are not equal.[22] The pulse sequence was optimised to suppress decoherence for a system of spins coupling to a non-interacting boson bath. It has been shown that the Uhrig scheme is independent of the model employed.[23] In other words, it can be applied to an arbitrary dephasing Hamiltonian as for example a spin bath model where the coherence dephasing of the observed spin originates from the bath-induced magnetic field fluctuations due to intrabath interactions. This particular spin bath model corresponds to the situation encountered in spin-labelled macromolecules with surrounding matrix nuclei.

For the Uhrig dynamical decoupling sequence of total sequence time T with n pulses, the $n+1$ interpulse delays τ_j with $1 \leq j \leq n + 1$ can be computed from[23]

$$\tau_j = \frac{1}{2} \left[\cos \left(\frac{\pi(j-1)}{n+1} \right) - \cos \left(\frac{\pi j}{n+1} \right) \right] T \tag{35}$$

which results in a sequence being symmetric with $\tau_{n+2-j} = \tau_j$. Starting with a $\pi/2$ pulse and setting $n = 2$, the Uhrig sequence is equivalent to the CP cycle. Upon addition of a pump sequence this setup recovers the 5-DEER experiment. For larger n, Uhrig schemes deviate from CP schemes.

These DD schemes can be used as DEER observer sequences for either determining the relaxation behaviour of the refocused spin echo at the acquisition position or alternatively incorperated in the corresponding DEER experiment. Both relaxation and DEER setups can be classified by the order n of underlying dynamical decoupling observer scheme: A pulse sequence of $(\pi/2 - (\tau/n - \pi - \tau/n)_n$-echo acquisition) is refered to as DD-DEER scheme of order n. The relaxation pulse sequences are displayed in Fig. 5. Hence, the 5-DEER experiment is of order 2 and 7-DEER is described by $n = 3$. The order n of the DD observer scheme corresponds to the number of required pulses at the pump frequency.

2 Material and Methods

2.1 Model systems

Model systems were chosen to mimic typical DEER application scenarios for which the relaxation behaviour of the nitroxide spin in the different spin environments (i.e. protonated, partly or completely deuterated solvent) was studied.

TEMPOL served as a model system to study the direct solvent influence on the relaxation behaviour of the nitroxide spin. For this system two different concentrations ($\approx$ 10 μM and 100 μM) which lie within the typical DEER concentration range were used to examine the effect of sample concentration on relaxation.

The double mutant T4 Lysozyme-72-131 represents the class of soluble proteins with a spin environment consisting of solvent as well as protein.

WALP23 is a synthesised α-helical trans-membrane peptide which was chosen as a model for proteins in a lipid membrane.

The rigid biradial MSA236 served as a sample with a well defined distance due to minimised conformational flexibility which allowed for DEER calibration measurements. Its chemical structure is shown in Fig. 15b on page 55. The relaxation behaviour of this system was approximated by a structurally similiar rigid monoradical MS 107 as shown in Fig. 4. Relaxation traces measured on the biradical would otherwise display dipolar modulation which would complicate the extraction of the relaxation law.

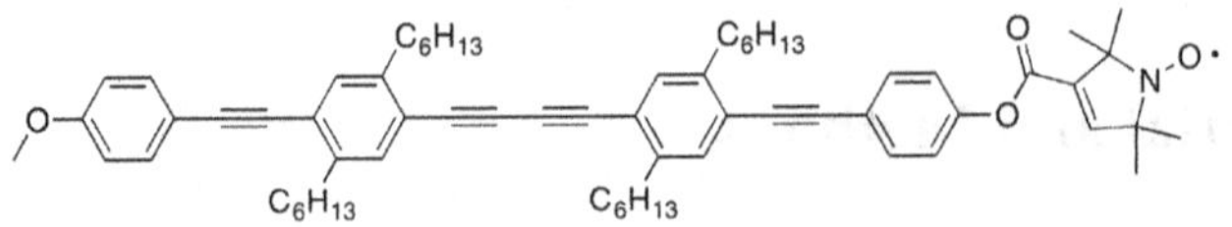

Fig. 4: Chemical structure of monoradical MS 107 with close structual resemblance to the biradical MSA 236. This allows to mimic the relaxation behaviour of the biradical.

- TEMPOL in
    - H_2O-Glycerol (HGly)
    - H_2O-deuterated Glycerol (d_8Gly)
    - D_2O-d_8Gly

- T4 Lysozyme-72-131 in
    - H_2O-HGly
    - H_2O-d_8Gly

- WALP23-2 in
    - H_2O
    - D_2O

- WALP23-11 in
    - H_2O
    - H_2O-d_8Gly

- Rigid Monoradical MS 107 in deuterated o-terphenyl (d_8-OTP)

- Rigid Biradical MSA 236 in d_8-OTP

2.1.1 Sample preparation

1:1 volume mixtures of HGly or d_8Gly and H_2O or D_2O were used to avoid crystallisation upon freezing TEMPOL or T4 Lysozyme samples in liquid nitrogen.

2.1.1.1 TEMPOL

10 μM and 100 μM TEMPOL solutions were prepared from a 1 mM TEMPOL stock solution in DMSO by diluting to 20 μM and 200 μM with H_2O or D_2O and addition of HGly or d_8Gly. The exact spin concentration was measured using continuous wave (CW) EPR relative to a reference sample of known spin concentration.

2.1.1.2 Rigid radicals MS 107 and MSA 236

The synthesis of the monoradical MS 107 and biradical MSA 236 was carried out by Professor A. Godt at Bielefeld University, Germany. A solid 50 μM mixture of either sample with d_8-OTP was placed into a sample tube and melted with a heat gun at 70 °C, a glassy frozen sample was obtained by subsequent shock-freezing in liquid nitrogen.

2.1.1.3 T4 Lysozyme

A T4 Lysozyme double mutant with cystein residues included at sites 72C and 131C was spin labelled with (1-Oxyl-2,2,5,5-tetramethylpyrroline-3-methyl) methanethiosulfonate (MTSSL). The spin labelling protocoll involved breaking of the S-S bridges via reduction using 20 μL of a freshly prepared 100 mM butan-2,3-diol-1,4-dithiol (DTT) solution for 2 mL cooled protein. The sample was incubated for two hours at 4°C. Subsequently, DTT was removed with a PD_{10} column using a gravity protocol and buffer pH set to 7.6 (40 mM NaCl, 25 mM 3-morpholinopropane-1-sulfonic acid (MOPS), 10 % H-Gly). For the labelling step two-fold label concentration per site relative to the protein concentration was used. The MTSSL solution was added immediately to the eluate from the column and was incubated for another 30 minutes while agitated at room temperature. The remaining free label was removed by a PD_{10} column and 10 kDa centricons were used for concentrating the sample. The final protein concentration was obtained from UV absorbance at 280 nm whereas the spin concentration was measured by CW EPR relative to a standard TEMPOL sample of known concentration. The final sample was obtained by adding HGly or d_8Gly in equal amounts to the spin labelled Lysozyme resulting in a final spin concentration of ≈ 13 μM. The modulation depth of 0.454 in a 4-DEER trace acquired with the commercial Q-band spectrometer as described in section 2.2.2.2 indicates a labelling efficiency of > 95%.

2.1.1.4 WALP23

The single mutants WALP23-2 and WALP23-11 of nominal 50 μM were spin labelled using MTSSL as part of a semester project by Max Doppelbauer. Details on the spin labelling procedure and probe preparation are specified in [24] under 3.1 *'Sample Preparation'*.

2.2 Instrumentation

2.2.1 *X band*

For CW EPR measurements of the spin concentration of TEMPOL and T4 Lysozyme samples a Bruker E500 X-band spectrometer operated by Xepr 2.2b.27 on a Suse 8.2 Linux with a Bruker SHQ resonator was used.

All pulsed experiments at X-band frequencies were performed on a home-built X-band spectrometer in combination with an arbitrary waveform generator (AWG) further specified in [25] under 2.1 *'Instrumentation'*.

2.2.2 *Q band*

2.2.2.1 Home-built spectrometer

Relaxation measurements for all TEMPOL samples were carried out using the home-built Q-band spectrometer operated from a standard console (Bruker) and equipped with a pulsed traveling wave tube (TWT) amplifier. The output power amounts to 150 W. The home-built TE102 rectangular resonator allowed for usage of oversized tubes of 3.0 mm diameter.[15]

2.2.2.2 Commercial

All Q-band DEER experiments as well as relaxation measurements (apart from TEMPOL samples) were performed on a commercial Bruker Elexsys E580 Q-band spectrometer with an output power of 200 W. Shaped pump pulses were generated with an Agilent M8190A 12 GS/s AWG. A more detailed description of the instrumentation can be found in[26] under 3.2 *'Instrumentation'*.

2.3 Experiments

All pulsed EPR measurements were performed at 50 K with $\approx$ 40 µl sample volume in quartz tubes of $\approx$ 3.0 mm outer diameter. The integration gate at the acquisition position was centered around the echo maximum whereas the width was set to twice the observer pulse length (24 ns). For offset-free signal detection a [(+x) - (-x)] phase cycle on every initial $\pi/2$ pulse of the pulse sequences was carried out.

For concentration measurements using CW EPR 2 µl sample volume was filled in glass capillaries.

2.4 Relaxation Measurements

Relaxation measurements with the observer sequence of the corresponding 4, 5-, 7- and 9-DEER (corresponding to order n = 1, 2, 3 and 4) and 7 and 9-Uhrig-DEER (n = 3 and 4) experiments were performed on TEMPOL, T4 Lysozyme, WALP23 and the monoradical MS 107 under the above specified conditions at Q band.

For MSA 236 and T4 Lysozyme, relaxation experiments were also carried out at X band prior to corresponding DEER experiments for 4-, 5- and 9-DEER scheme as well as the 9-Uhrig scheme. Where the setup in terms of delays and pulse duration corresponds to the respective DEER observer sequence as described below.

2.4.1 Pulse sequences

Fig. 5 displays the pulse sequences employed for the relaxation measurements with pulse lengths $t_{\pi/2} = t_\pi = 32$ ns at X band and $t_{\pi/2} = t_\pi = 12$ ns at Q band. All time delays after generation of the primary echo $(\pi/2 - \tau/n - \pi)$ were incremented by $\Delta t = 4$ ns, shifting the acquisition position accordingly. All relaxation traces displayed in this thesis such as in Fig. 11 have an x-axis corresponding to the absolute time. To allow for comparison of the echo intensity as a function of time the inital acquisition time for Δt = 0 was chosen identical for all relaxation pulse sequences.

The 4-DEER and 5-DEER relaxation sequence share the same number of pulses whereas the latter is characterised by symmetric pulse delays $(\tau/2 - \tau - \tau/2)$ as can be seen in Fig. 5. Due to imperfect pulses a stimulated echo $(\pi/2 - \tau/2 - \pi/2 - \tau/2 - \pi/2)$ additionally to the desired refocused echo $(\pi/2 - \tau/2 - \pi - \tau/2 - \pi)$ arises at the 5-DEER-acquisition position. This complication does not occur in the 4-DEER setup as depicted in Fig. 6. To measure the relaxation of the pure refocused echo the stimulated and refocused echo needed to be separated in time by shifting the second π pulse by the delay $\delta\tau$ as illustrated in Fig. 6b. Delay $\delta\tau$ was chosen such that it only corresponds to a fraction of the total sequence time. This allows to neglect any relaxation occuring during the additional $\delta\tau$ time. Setting $\delta\tau = 0$ in a relaxation measurements leads to an underestimation of the real relaxation time as the stimulated echo signal contribution decays with approximately T_1.

Instead of the introduced time shift, stimulated and refocused echo could be alternatively separated by phase cycling. Whereas this approach is feasible for the 5-DEER sequence, the number of required cycles quickly increases with additional π pulses. More specifically two π refocusing pulses require a four-step phase cycle, such that

30

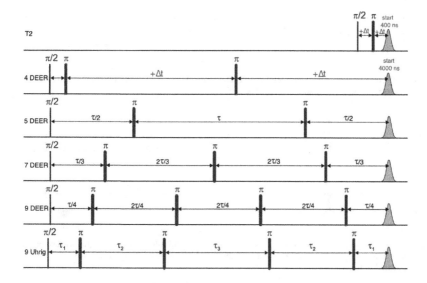

Fig. 5: Pulse sequences for T_2 relaxation measurements including the standard primary echo setup as well as DEER observer frequency derived sequences. The basis τ for the 4, 5, 7- and 9 DEER sequences were chosen so that the intial distance between the first pulse in the sequence and the echo is identical in all measurements. The time delays τ_1, τ_2 and τ_3 for the 7 and 9-Uhrig sequence were obtained from equation (35) for $T = t_{acq}$. 7-Uhrig is not displayed here.

for a number m of π refocusing pulses in the observer sequence a 2^m step phase cycle is required. This ignificantly increases the number of necessary acquisitions per experiment.

Therefore a time separation represents the more elegant approach and was used for all DEER sequences of order $n = 1, 2, 3$ and 4 and Uhrig-DEER sequences of order $n = 3$ and 4. Increasing the number of π pulses does not only lead to different stimulated echos but additionally creates a number of refocused echos. For the 7-DEER sequence already four different echos were observed at the proximity of the acquisition position and care had to be taken to select the correct echo for signal acquisition. For the DEER relaxation measurements crossing echos for all sequences of order $n > 2$ were observed,

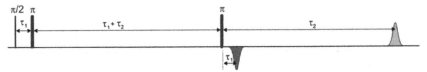

(a) 4-DEER observer sequence with stimulated echo (dark) at $3\tau_1 + \tau_2$ naturally separated in time from the refocused echo (light) at $2\tau_1 + 2\tau_2$ of opposite phase.

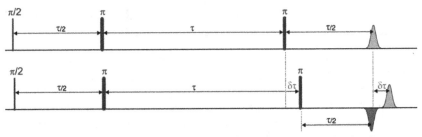

(b) 5-DEER observer sequence overlap of stimulated (dark) and refocused echo (light) with opposite phase at the acquisition position. Upon shifting the last π pulse by delay $\delta\tau$ the two echos are separated in time.

Fig. 6: Stimulated and refocused echo positions in 4- and 5 DEER observer scheme.

such that a negative time shift $\delta\tau$ was employed to all π pulses after the primary echo part of these sequences.

2.5 DEER Measurements

2.5.1 *Pulse sequence*

The 5-DEER scheme introduced by Borbat, Georgieva and Freed is displayed in Fig. 3c and shows the figure from the original publication.[4] This sequence is recovered upon adding a pump pulse sequence at ω_B to the stationary $(\pi/2 - \tau/2 - \pi - \tau - \pi - \tau/2)$ sequence as displayed in Fig. 6b ($\delta\tau = 0$).

In section 2.4.1 complications arising from overlapping stimulated and refocused echo for relaxation measurements have been discussed. Fig. 7 illustrates the effect for DEER measurements as the unshifted 5-DEER scheme leads to a reduction in modulation depth Δ. This effect is due to the unmodulated stimulated echo signal contribution at the acquisition time. Note that the 5-DEER publication by Borbat, Georgieva and Freed does not discuss this issue. [4]

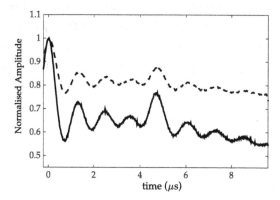

Fig. 7: Rect-5'-DEER traces for MSA 236 at Q band obtained with the 5'-DEER sequence as shown in Fig. 6b with setting $\delta\tau = 0$ ns (‑ ‑) where stimulated and refocused echo overlap at the acqusition position. With $\delta\tau = 72$ ns (—) the unmodulated stimulated echo does not contribute to the signal at the acqusition position any longer. A significant increase in modulation depth Δ is visible.

Implementation of the original 5-DEER scheme was not possible due to restrictions in the Bruker PulseSpel Software which generally does not allow negative time increments. In other words, pushing the pump pulse 3 from right to left as shown in Fig. 3c could not be realised. Hence, the sequence was adjusted to a 5'-DEER scheme such that the time delay t was incremented and the stationary pump pulse was moved to position 5' as depicted in Fig. 8a. The dashed sequence nomenclature referes in the following to DEER experiments with pump pulses at the just specified position. In contrast, the dashed nomenclature does not apply to the standard 4-DEER experiment nor to higher order relaxation setups. Additionally to the 5' pump pulse, the second observer π pulse was shifted to obtain a pure refocused echo at $(2\tau + 2\delta\tau)$. The dipolar evolution time for the modified 5'-DEER scheme can be obtained as described in section 1.4.3.1 and amounts to $t_{\text{dip}}^{5'\text{-DEER}} = 2\,(t - \delta T)$ such that the signal is modulated with $\cos\left[d\,(t - \delta T)\right]$

which is identical to the 5-DEER signal modulation defined in equation (30). The time origin of the dipolar evolution occurs for $t = \delta T$ in both 5-pulse schemes, however the 5′ setup reaches this point earlier in time. This is experimentally more convenient and results in a reversal of the modulated time trace relative to the Borbat, Georgieva and Freed 5-DEER setup. To obtain a deadtime-free 5′-DEER sequence, δT has to be chosen larger than the mininum delay t_{min} between a pump and an observer pulse. This experimental condition applies to all higher order CP derived DEER sequences.

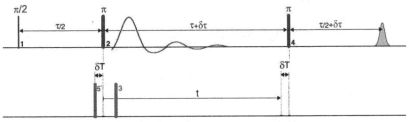

(a) 5′-DEER scheme with reversed modulated time trace relative to 5-DEER as shown in Fig. 3c.

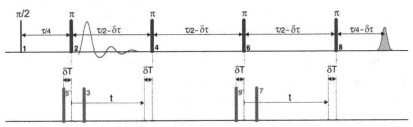

(b) Extension of 5′-DEER to a 9′-DEER experiment which corresponds to $N = 2$ for the scheme displayed in Fig. 3c with adjustments analog to the 5′-DEER sequence. As indicated the dipolar time decreases whereas the signal oscillates with doubled frequency with respect to the 5′-DEER signal if the same pump pulse increment is applied to both pulses 3 and 7.

Fig. 8: 5′-DEER and 9′-DEER experiments with fixed interpulse delays in the observer sequence (with pulses 1, 2, 4 etc.) at ω_A which also applies to the delay δT in the pump sequence (with pulses 5′, 3 etc.) at ω_B. The relative position of pump pulse 3 (and 7) is moved by incrementing t. A deadtime-free sequence is obtained for $\delta T > t_{min}$ with t_{min} being the minimum delay between a pump and an observer pulse to prevent pulse overlap artifacts. The numbering of the pulses follows the original scheme as displayed in Fig. 3a.

DEER experiments of even order n were carried out which are analogue to the DEER relaxation setups and require addition of an even number n pump pulses. This is an advantage in case of adiabatic chirp pump pulses as recognised by Spindler.[5] The frequency sweep, characteristic for the chirp pulse, leads to a dispersion in terms of pump spin inversion times. For an even number of chirp pulses this dispersion can be refocused by applying chirp pairs with opposite sweep direction. Because this refocusig is incompelete for an odd number of pump pulses the 7-DEER scheme was not considered.

Therefore 9-DEER corresponds to the next higher order of dynamical decoupling and is characterised by $n = 4$. This experiment can be obtained by simply repeating the 5-DEER scheme as already suggested by Borbat, Georgieva and Freed.[4] The symmetrised 9-DEER scheme suffers from the same stimulated echo complications as the 5-DEER setup, so that the $n = 4$ DEER experiment was implemented as 9'-DEER experiment shown in Fig. 8b. The observer spin echo refocuses at $(2\tau - 4\delta\tau)$ with $t_{\text{dip}}^{9'\text{-DEER}} = 4\,(t - \delta T)$, such that the amplitude is modulated by $\cos\left[2d\,(t - \delta T)\right]$. The upper bound of both variable pump pulse delays is $t \leq \tau/2 - \delta\tau - \delta T$, leading to a maximum dipolar evolution time of $t_{\text{dip, max}}^{9'\text{-DEER}} = 2\tau - 4\delta\tau - 8\delta T$ relative to $t_{\text{dip, max}}^{5'\text{-DEER}} = 2\tau + 4\delta\tau - 2\delta T$. Due to the opposing sign of $\delta\tau$ for the 5'-DEER and 9'-DEER setup, the available dipolar evolution time $t_{\text{dip, max}}^{5'\text{-DEER}}$ appears to be longer than $t_{\text{dip, max}}^{9'\text{-DEER}}$ for the same choice of $\tau, \delta\tau$ and δT. However, as $\delta\tau$ and δT are set to much shorter lengths relative to τ, the gain in maximum dipolar evolution time for the 5'-DEER sequence is only small. However the 9'-DEER scheme has an advantage for big enough ξ values such that the refocusing allows to choose longer τ values for the 9'-DEER experiment relative to the 5'-DEER setup.

In analogy to equation (30) on page 18 the intramolecular echo amplitude modulation function for the 9'-DEER setup $V^{9'\text{-DEER}}(t)$ can be derived. This expression is made up of 2^4 terms. However as only the contributions that depend on t are relevant for the 9'-DEER trace, signal contributions arising from $(q_3 p_{5'} p_7 p_{9'})$, $(p_3 q_{5'} p_7 q_{9'})$, $(q_3 p_{5'} q_7 p_{9'})$, $(q_3 p_{5'} q_7 p_{9'})$, $(q_3 p_{5'} q_7 q_{9'})$ and $(q_3 q_{5'} q_7 q_{9'})$ are not listed. The angular brackets denote again the average over all p_k, q_k and interspin vector orientations hidden in d.

$$
\begin{aligned}
V^{9'\text{-DEER}}(t) \propto &\langle (p_3 p_{5'} p_7 p_{9'}) \cos\left[2d\left(t - \delta T\right)\right] + (p_3 q_{5'} p_7 p_{9'}) \cos\left[2d\left(t - \delta T/2 - \tau/8\right)\right] \\
&+ (q_3 p_{5'} p_7 p_{9'}) \cos\left[d\left(t - \tau/4\right)\right] + (p_3 p_{5'} q_7 p_{9'}) \cos\left[d\left(t - 2\delta T + \tau/4\right)\right] \\
&+ (p_3 q_{5'} q_7 p_{9'}) \cos\left[d\left(t + \delta\tau - \delta T\right)\right] + (q_3 p_{5'} p_7 q_{9'}) \cos\left[d\left(t - \delta\tau - \delta T\right)\right] \\
&+ (q_3 q_{5'} p_7 p_{9'} + p_3 p_{5'} q_7 q_{9'}) \cos\left[d\left(t - \delta T\right)\right] + (q_3 q_{5'} p_7 q_{9'} + p_3 q_{5'} q_7 q_{9'}) \cos\left[d\left(t - \tau/4\right)\right] \rangle
\end{aligned}
\tag{36}
$$

Expression (36) can be simplified by two assumptions: First, equal spin inversion probabilities for each pump pulse (3, 5′, 7 and 9′) are assumed. This assumption is reasonable if pump pulses are well reproducible once the respective pulse parameters have been set. Secondly as $p \ll q$ holds for optimised pump pulses, terms of low probability in equation (36) can be neglected, leading to

$$
\begin{aligned}
V^{9'\text{-DEER}}(t) \propto &\langle p^4 \cos\left[2d\left(t - \delta T\right)\right] + p^3(1-p) \cos\left[2d\left(t - \delta T/2 - \tau/8\right)\right] \\
&+ p^3(1-p) \cos\left[d\left(t - \tau/4\right)\right] + p^3(1-p) \cos\left[d\left(t - 2\delta T + \tau/4\right)\right] \rangle
\end{aligned}
\tag{37}
$$

where the relation $q = (1 - p)$ was used. Here the angular brackets denote the average over all interspin vector orientations hidden in d. As $\frac{p^4}{p^3(1-p)} = \frac{p}{(1-p)}$ applies, the signal contributions weighted by $p^3(1 - p)$ in equation (37) are expected to scale comparable to the symmetric 4-DEER artifact in the 5-DEER signal for identical pump pulses. Therefore, additionally to the main 9′-DEER signal contribution three additional terms have to be considered corresponding to $(p_3 q_{5'} p_7 p_{9'})$, $(p_3 p_{5'} p_7 q_{9'})$ and $(p_3 p_{5'} q_7 p_{9'})$. In expression (37) the first $p^3(1 - p)$ term refocuses at $t = \delta T/2 + \tau/8$. As $\tau \gg \delta T$ the two signal contributions are well separated in time. The third and fourth term in equation (37) are only modulated with half the dipolar frequencies as either one of the moving pump pulses enter with q into the expression. For the above specified experimental conditions both terms refocus at similiar dipolar time t. Hence the $(q_3 p_{5'} p_7 p_{9'})$ and $(p_3 p_{5'} q_7 p_{9'})$ terms coincide in time and can be vieved as the analogue to the $p_3 q_5$ signal contribution in the 5′-DEER setup.

In modification of the CP inspired multipulse DEER setups the observer sequence may also be modified to a Uhrig decoupling scheme. For order $n = 2$ this coincides with the symmetric CP pulse arrangement as used in 5-DEER. A possible 7-Uhrig-DEER sequence was not considered due to the uneven number of pump pulses. Therefore a 9'-Uhrig-DEER experiment as depicted in Fig. 9 allows for the implementation of this asymmetric DD observer scheme.

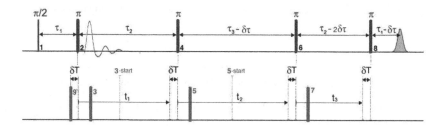

Fig. 9: 9'-Uhrig-DEER sequence with interpulse delays τ_1, τ_2 and τ_3 defined by equation (35) on page 24 for a total sequence length of T. Relative to the 9'-DEER setup the position of the first π pump pulse 9' is kept fixed at a delay δT whereas pump pulses 3, 5 and 7 are pushed with identical increments ($t = t_1 = t_2 = t_3$) so that the smallest interpulse delay between observer pulses 6 and 8 determines the available dipolar evolution time. For t increments identical in size to the 5 pulse scheme, the signal oscillates with three times the frequency with respect to 5'-DEER modulation for identical incrementation of t. The time shift $\delta\tau$ reduces the available time window, yet allows for a stimulated-echo-free acqusition position. Setting $t_{3,\,min}$ and $t_{5,\,min}$ to the dashed starting position labelled 3- and 5-start, allows to reduce t_{min}.

Similiar to the derivation of equation (37) for the 9'-DEER sequence, only signal contribution with weight p^4 and $p^3(1 - p)$ are discussed in the following for the 9'-Uhrig sequence. The expression for the 9'-Uhrig intramolecular echo amplitude modulation function $V^{9'Uhrig}(t)$, specified in equation (38), is independent of the interpulse delay τ_3. This delay can be eliminated by the refocusing condition $\tau_3 = 2(\tau_2 - \tau_1)$.

$$V^{9'\text{-Uhrig}}(t) \propto \langle (p_3 p_{5'} p_7 p_{9'}) \cos\left[d\left(2\tau_1 - 2\tau_2 + \delta T + \delta\tau + (t_1 + t_2 + t_3)\right)\right]$$
$$+ (p_3 p_{5'} p_7 q_{9'}) \cos\left[d\left(2\tau_2 - \tau_1 - \delta\tau - (t_1 + t_2 + t_3)\right)\right]$$
$$+ (q_3 p_{5'} p_7 p_{9'}) \cos\left[d\left(2\tau_2 - \tau_1 - \delta T - \delta\tau - (t_2 + t_3)\right)\right] \tag{38}$$
$$+ (p_3 q_{5'} p_7 p_{9'}) \cos\left[d\left(\tau_2 - \tau_1 - \delta T - \delta\tau + (t_1 - t_3)\right)\right]]$$
$$+ (p_3 p_{5'} q_7 p_{9'}) \cos\left[d\left(\tau_1 - \tau_2 - \delta T + t_3\right)\right]]\rangle$$

The convenient choice of $t = t_1 = t_2 = t_3$, uses the same time increment t for all pump pulses 3, 5 and 7. Applying the same assumptions as used for simplying equation (36) to (37), leads to

$$V^{9'\text{-Uhrig}}(t) \propto \langle p^4 \cos\left[d\left(2\tau_1 - 2\tau_2 + \delta T + \delta\tau + 3t\right)\right] + p^3(1-p) \cos\left[d\left(2\tau_2 - \tau_1 - \delta\tau - 3t\right)\right]$$
$$+ p^3(1-p) \cos\left[d\left(2\tau_2 - \tau_1 - \delta T - \delta\tau - 2t\right)\right] + p^3(1-p) \cos\left[d\left(\tau_1 - \tau_2 - \delta T + t\right)\right]]\rangle \tag{39}$$

Note, how the fourth signal contribution in equation (38) $(p_3 q_{5'} p_7 p_{9'})$ vanishes for $t_1 = t_3$. Considering the main signal modulation with weight p^4, the minimum dipolar evolution time is reached at $t_{1,\min} = t_{2,\min} = t_{3,\min} = \delta T$ and corresponds to $t_{\text{dip, min}}^{9'\text{-Uhrig}} = 2\tau_1 - 2\tau_2 + 4\delta T + \delta\tau$. For $t_{\text{dip, min}}^{9'\text{-Uhrig}} < 0$ the time origin is sampled, a condition which is fulfilled for the 9'-Uhrig setup due to two reason. First of all, δT and $\delta\tau$ are both chosen to be much shorter than any Uhrig interpulse delay τ_j. Secondly, $\tau_2 > \tau_1$ holds according to equation (35) on page 24. The smallest interpulse delay between the observer pulses 6 and 8 determines the the maximal time window over which t can be incremented, as illustrated in Fig. 9. Instead of starting with $t_{1,\min} = t_{2,\min} = t_{3,\min} = \delta T$, the intial delays $t_{1,0}$ and $t_{2,0}$ can be set to $t_{1,0} = \delta T + \delta\tau$ and $t_{2,0} = \delta T + \delta\tau + (\tau_3 - \tau_2)$. Using again the refocusing condition, this setup yields $t_{\text{dip, min'}}^{9'\text{-Uhrig}} = 6\delta T + \delta\tau - \tau_2$ and corresponds to starting position of pump pulses 3 and 5 indicated by the dashed lines labelled 3- and 5-start respectively in Fig. 9. This allows to reduce the fraction of dipolar evolution observed before the time origin as $|t_{\text{dip, min'}}^{9'\text{-Uhrig}}| < |t_{\text{dip, min}}^{9'-Uhrig}|$ applies.

Moreover equation (39) shows three additonal signal contribution, each weighted by

$p^3(1 - p)$. The $(p_3p_5{}'p_7q_9{}')$ contribution is modulated with the same frequeny as the desired p^4 weighted signal, but is separated in time by $\tau_1 + \delta T$. Signal artifacts arising from $(q_3p_5{}'p_7p_9{}')$ and $(p_3p_5{}'q_7p_9{}')$ can also be distinguished from the main signal as they are only modulated by two thirds or a third of the main dipolar frequency.

Performing DEER experiments on a pair of nitroxide labels as observer and pump spin requires the excitation of two distinct regions of their nearly identical absorption spectrum. The nitroxide absorption spectrum differs for X- and Q-band frequencies, 9.5 GHz and 34.5 GHz respectively. This implies different pulse lengths and frequency offsets between pump and observer frequencies for artifact free DEER traces of optimal sensitivity.[15] The DEER setup at the two frequencies is specified in the following.

2.5.1.1   X band

For X-band frequencies a 12-65-32 excitation scheme was used which applies pump pulses of 12 ns length on the central maximum of the spectrum. The observer frequency was placed at lower fields with a frequency offset of 65 MHz relative to the pump frequency. The observer pulses were adjusted to 32 ns for both $\pi/2$ and π pulses.[15]

2.5.1.2   Q band

The nitroxide absorption spectrum at Q band is broader and more asymmetric with respect to X band. This allows for harder observer pulses of 12 ns lengths while choosing a bigger frequency offset of 100 MHz between the pump and the observer frequencies. Pump pulses were set to 12 ns as well, leading to a 12-100-12 excitation scheme.[15] All pulses were adjusted by nutation experiments.

2.5.2 Pump pulses

Non-rectangular pump pulses were used to minimise the signal contribution from inverting pump spins once, but less than two or four times for 5'-DEER and 9'-DEER experiments, respectively. For the latter two experiments pump pulses were organised in pairs such that the frequency sweep was carried out in opposite directions in order to refocus pump spin inversion dispersion. Shaped adiabatic pump pulses were first tested on the model system MSA 236 and subsequently applied to T4 Lysozyme in H_2O-HGly.

2.5.2.1 Design

The optimised non-rectangular pump pulses were pre-simulated by a graphical user interface (GUI) programmed in Matlab by Andreas Dounas.[11].

The programme allows to incorporate the resonator profile which was determined experimentally as described in subsection 2.5.3. This allowed to generate resonator compensated pump pulses which have been shown to enhance inversion performance.[27] The convolution of the experimental resonator profile with the amplitude and frequency modulation function of the simulated pump pulse in the frequency domain gave a visual impression on the expected inversion quality. Furthermore, the GUI output nominal values for adiabaticity Q_{min} and inversion I, as defined by equation (20) and (25), were used to find optimal pump inversion pulses for each experimental setup. Each pulse is defined by the central frequency, the sweep bandwidth $\Delta\nu$ both in GHz and the pulse length t_p in ns. HSh pulses required as specification the order h and the parameter β_{HS}. For asymmetric HSh pulses h_l and h_r defined the left and right order for the respective pulse flank. Generally, optimal inversion performance can be achieved by steep edges and a flat pulse plateau in the frequency domain.

2.5.3 Resonator compensation

The direct relation between the nutation frequency of a resonant spin and the magnetic field strength for a given frequency f can be used to obtain the resonator profile $\nu_1(f)$ based on nutation experiments as described in section 1.4.2. Pulse lengths for X- and Q-band resonator profile characterisation were set to 16 ns ($\pi/2$) - 32 ns (π) and 12 ns ($\pi/2$) - 24 ns (π).[27]

At Q band an RLC-resonator profile was fitted to the experimental profile to obtain parameters for resonator compensation incorporated into the pump pulse design. At X band, experimental $\nu_1(f)$ profiles were used for the pulse compensation.

Secondly, a non-linear calibration $\nu_1(a_s)$ of the digital pulse amplitude a_s was carried out as described in[27] to compensate for amplitude imperfections introduced by the resonator profile and the deviation created in the TWT relative to the digital pulse.

2.5.3.1 Inversion characterisation

The inversion efficiency I as defined in equation (25) was used to characterise the pump pulses of various pulse shapes as a function of frequency (rectangular, linear chirp and HSh pulses). For this a Hahn echo scheme was used which is displayed in Fig. 1a.[27] At Q band the pulse lengths were set to 12/24 ns whereas X-Band requires pulse lengths of 16/32 ns. The echo was integrated over 24 and 32 ns, respectively, by setting the observation frequency to the intended pump frequency in the DEER experiment. At X band, the excitation profile of a pulse was obtained by simultaneously stepping the observation frequency with the field. For monochromatic pump pulses, ν_1 corresponded to the observation frequency. For the linear chirp and HSh pulses the center frequency was stepped. At Q band only the observation frequency was incremented.
A phase cycle [+(+x) - (-x)] on the first $\pi/2$ pulse allowed for offset cancellation.

2.5.4 DEER signal analysis

For standard DEER signal analysis DeerAnalysis[28] was used. The procedure is summarised in section 1.4.4.

3 Results and Discussion

3.1 Relaxation

Relaxation measurements for all samples were carried out. The resulting relaxation traces for TEMPOL, T4 Lysozyme, WALP23 and MS107 at Q-band frequencies were normalised by the maximum signal intensity of the 4-DEER trace. This normalisation allows to compare the refocusing scheme performance relative to the 4-DEER setup as a function of time. X-band traces are displayed in absolute echo integrals. Time zero in all relaxation figures corresponds to about 4.2 µs echo time with discrepencies due to $\delta\tau > 0$ for 5-DEER pulse sequences, compare Fig. 6 on page 32, and $\delta\tau < 0$ for 7- and 9-DEER relaxation experiments.

The 4-DEER traces were fitted by a single stretched exponential relaxation law specified in equation (33). Representative data for the fits is shown in Fig. 11 on page 46 for TEMPOL in H_2O-HGly. The fit routine relied on a least-squares criterion and was carried using Matlab. The obtained fitting parameters T_m and ξ allow to characterise the relaxation process.

3.1.1 TEMPOL

Tab. 1 summarises the effect of spin concentration and solvent on the TEMPOL relaxation behaviour.

Higher spin concentration leads to quicker echo decay as indicated by a shorter phase relaxation time T_m and a smaller stretch parameter ξ. Only for 282 µM TEMPOL in D_2O-d_8Gly ξ increases relative to the lower concentrated sample. This apparent decrease in relaxation rate at long times is offset by the strong decrease for T_m so that the higher

Tab. 1: Fit parameter T_m and ξ for TEMPOL 4-DEER relaxation measurements.

conc. [μM]	Solvent (1:1)	T_m [μs]	ξ
11	H_2O-HGly	1.95	1.29
>75	H_2O-HGly	1.48	1.16
NA	H_2O-d_8Gly	2.77	1.27
<128	H_2O-d_8Gly	2.54	1.12
44	D_2O-d_8Gly	9.96	0.72
282	D_2O-d_8Gly	1.65	0.92

concentrated probe shows an overall quicker relaxation behaviour in comparison to the 44 μM sample. However, $\xi < 1$ for TEMPOL in D_2O-d_8Gly deviates from an expected stretch parameter of $\xi \geq 1$.

Apart from this deviation, higher concentrated samples show quicker relaxation, hence, at concentrations >75 μM spin-spin interaction represents the dominating relaxation mechanism. Instead, for the lower concentration regime of $\approx$ 10 μM, somewhat larger ξ are found. This is closer to Gaussian type decay which is typical for diffusion controlled relaxation. Full or partial substitution of solvent protons by deuterons leads to an overall decrease in relaxation rate. This effect is already noticable when using deuterated glycerol in combination with H_2O.

Literature on electron spin echo dephasing for nitroxyl radicals in glassy solvents report ξ values between 2.0 and 2.3 for 2,2,6,6-tetramethyl-4-oxo-1-piperidinyloxy (TEMPONE) in solvents without methyl groups.[29] More specifically for TEMPONE (100 - 400 μM, X Band) at 40 K in a 1:1 mixture of H_2O-Glycerol the spin echo decay was characterised by T_m = 4.6 μs and ξ = 2.3.[29]

Despite the structual similiarities of TEMPONE and TEMPOL, ξ values of about 2 where not observed for the DEER-T_2 measurements. This deviation from literature can be explained by the inherent nature of the relaxation experiment. Assuming a Gaussian type decay i.e. ξ = 2 and an arbitrary phase relaxation time T_m of 4.6 μs, Fig. 10 shows the dependence of the apparent relaxation law on the acquisition time in the experiment. For the later acquisition time of 4.1 μs in the 4-DEER relaxation experiment, lower ξ and T_m values are obtained relative to a standard T_2 setup. From Fig. 10 it can understood that a Gaussian decay yields a slower relaxation law, characterised by ξ = 1.22 and T_m = 1.92 µs, if the echo time of the 4-DEER experiment is treated as the beginning of the relaxation process.

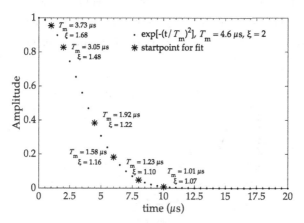

Fig. 10: Illustration of how the acquisition time in a T_2 relaxation experiment influences the relaxation parameter T_m and ξ. For a Gaussian type relaxation (ξ = 2, T_m = 4.6 µs), stretched exponential fitting yields different values for ξ and T_m. For later acquisition time, i.e. later starting points of the fitting routine, indicated by the stars, decreasing ξ and T_m values are obtained. Hence, a slower apparent relaxation behaviour is observed with increasingly later acquisition time.

To support this hypothesis by experiment, T_2 was measured using a standard refocused echo setup for TEMPOL relaxation in H_2O-HGly at 11 and >75 µM spin concentration, additionally to DEER-T_2 measurements. Fig. 11 and Tab. 2 show that earlier refocusing of the echo leads to a more Gaussian type relaxation behaviour for both high and low TEMPOL concentration.

Moreover, the fitting routine according to equation (33) was adopted so that the echo amplitude at $t = 0$ $V(0)$ served as the third fitting parameter. In this case the time t represents the absolute echo time after the initial spin excitation with associated fitting parameters T_m' and ξ'. This treatment shifts the time axis by the acquisition time and hence allows to extrapolate the relaxation process to the time directly after the first exciation. In advance, it was found that the time shifted fitting routine for the decay shown in Fig. 10 yields the correct values for T_m'' and ξ' up to an acquisition time of 5 µs. This time window includes the 4-DEER setup and hence the time-shifted fitting routine was applied to the 4-DEER trace of > 75 µM TEMPOL in H_2O-HGly which yielded $T_m'' = 3.95$ and $\xi' = 2.08$. This is in much better agreement with TEMPONE reference relaxation data and resolves the discrepancy of experimental finding from literature values.

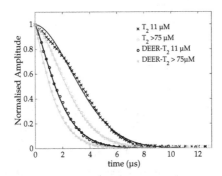

Fig. 11: Comparison of T_2 and DEER-T_2 relaxation measurements of 11 and >75 µM TEM-POL in H_2O-HGly. The solid lines represent the stretch exponential fit according to equation (33) with T_m and ξ parameter specified in Tab. 2.

Tab. 2: Fit parameter T_m and ξ TEMPOL in 1:1 H_2O-HGly relaxation measurements as displayed in Fig. 11.

T_m [µs] / ξ	11 µM	>75 µM
T_2	3.9944 / 1.82	2.9079 / 1.58
DEER T_2	1.9445 / 1.29	1.4778 / 1.16

3.1.2 T4 Lysozyme

The double mutant T4 Lysozyme spin labelled at position 72 and 131 with a spin concentration of ≈ 13 µM shows identical ξ values of 1.23 for both H_2O-HGly and H_2O-d_8Gly solvent environments as summarised in Tab. 3. In view of the results shown in Fig. 10, this stretch exponent may be consistent with a Gaussian relaxation law, which in turn suggests a water proton dominated mechanism. The deuterons influence the phase memory time T_m and slow down the relaxation process. The relaxation measurements with T4 Lysozyme-72-131 probe the spin environment of two different sites as the sample is a double mutant. Hence, the obtained T_m and ξ values reflect the net behaviour of both spin positions. However, both position 72 and 131 are solvent-exposed α-helical sites. Therefore deviation for the net relaxation behaviour from a single mutant Lysozyme sample is considered to be negligible.

Tab. 3: Fit parameter T_m and ξ for T4 Lysozyme-72-131 4-DEER relaxation measurements.

Solvent	T_m [µs]	ξ
H$_2$O-HGly	1.78	1.23
H$_2$O-d$_8$Gly	2.60	1.23

3.1.3 WALP23

The relaxation behaviour of the α-helical trans-membrane model peptide WALP23 spin labelled at the peptide end (position 2) and in the middle of the peptide (position 11) was charaterised by T_2 and the DEER-T_2 experiments. All WALP23 samples corresponded to a nominal concentrations of 50 µM with relaxation fitting parameters is summarised in Tab. 4.

For the DEER-T_2 measurements position 11 shows a quicker relaxation relative to position 2. At the former position close to simple exponential relaxation were obtained, both for H$_2$O and H$_2$O-d$_8$Gly. This is indicative for a spin environment in which the solvent is not decisive for the relaxation behaviour. T_m as well as ξ values differ for position 2 and 11 so that both parameters characterise the different spin environments. Comparing DEER-T_2 to the apparent relaxation laws obtained from T_2 measurements show similiar ξ value, whereas the T_m values deviate and show a reversed trend for deuteration for position 2. The difference in absolute value of T_m for position 2 is reasonable due to the earlier acquisition time for the T_2 measurements as illustrated in Fig. 10. Yet the T_m discrepancy for position 11 in H$_2$O cannot be explained.

Tab. 4: Fit parameter T_m and ξ for WALP23 4-DEER relaxation measurements.

Conditions		T_2		DEER-T_2	
Position	Solvent	T_m [µs]	ξ	T_m [µs]	ξ
2	H_2O	1.74	1.05	1.53	1.12
2	D_2O	1.68	1.03	1.62	1.12
11	H_2O	1.10	1.04	1.28	1.08
11	H_2O-d_8Gly	NA	NA	1.40	1.06

3.1.4 Comparison of free TEMPOL, spin-labelled T4 Lysozyme and WALP23

Comparing the apparent relaxation laws obtained for the three model systems TEMPOL, T4 Lysozyme and WALP23, the decreasing water proton influence leads to decreasing ξ values. This corresponds to larger deviation from the expected Gaussian decay typical for diffusion dominated relaxation. 50 µM WALP23-11 in H_2O and 11 µM TEMPOL in H_2O-HGly represent two extreme situations for an exponential relaxation law ($\xi = 1.04$) and Gaussian type relaxation ($\xi = 1.82$) for T_2 measurements, respectively.

Fig. 12a illustrates how frequent refocusing for $\xi > 1$ can slow down the relaxation process leading to significantly higher T_m values for an increasing number of π pulses. In contrast for $\xi \approx 1$, the relaxation time cannot be prolonged significantly neither with order n CP nor with Uhrig decoupling schemes as shown in Fig. 12b.

Low concentrated T4 Lysozyme in H_2O-d_8Gly ($\xi = 1.23$) represents an intermediate case in comparison to the simple exponential and near Gaussian decay. This stretch parameter was obtained from the 4-DEER relaxation experiment which is displayed Fig. 13a together with higher order relaxation traces.

For all systems studied, the 5-DEER relaxation traces display the highest absolute intensities with up to a factor of about 2.8 relative to the 4-DEER signal for WALP23-11 in H_2O. The higher signal intensity is due to the CP-type pulse setup for the 5-DEER experiment as shown in Fig. 5. In contrast, the asymmetric 4-DEER relaxation sequence does not allow for optimal decoupling.

Considering the echo intensities as a function of interpulse delay, DD schemes proof

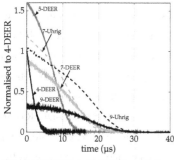

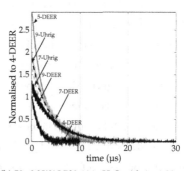

(a) 11 μM TEMPOL in H_2O-HGly with ξ = 1.82.

(b) 50 μM WALP23-11 in H_2O with ξ = 1.04.

Fig. 12: Echo relaxation traces obtained by different pulse sequences 4-DEER (▬), 5-DEER (▬), 7-DEER (▬), 9-DEER (▬), 7-Uhrig (⋅⋅) and 9-Uhrig (▬ ▬) at Q band. The amplitude is normalised to the 4-DEER amplitude. The traces show the effect of $\xi > 1$ upon multiple refocussing of the primary echo. The respective stretch parameters ξ are obtained from T_2 measurements.

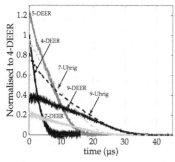

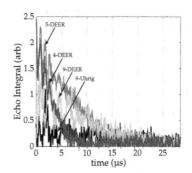

(a) Echo relaxation traces obtained by different pulse sequences 4-DEER (▬), 5'-DEER (▬), 7-DEER (▬), 9-DEER (▬), 7-Uhrig (⋅⋅) and 9-Uhrig (▬ ▬) at Q band.

(b) Echo relaxation traces obtained by different pulse sequences 4-DEER (▬), 5-DEER (▬), 9-DEER (▬) and 9-Uhrig (▬) at X band.

Fig. 13: Relaxation traces for T4 Lysozyme in H_2O-d_8Gly at Q and X band.

most successful for $\xi > 1$ as expected from theory with the 4-DEER setup showing the worst performance over all. Even for WALP23-11 in H_2O (ξ = 1.04) an intensity gain is achieved with the 5'-DEER experiment proofing best up to 5 µs. At this point in time the 4-DEER intensity has long decayed. For longer times higher order decoupling schemes show slightly higher intensities. The relative gain is much more distinct for TEMPOL in H_2O-HGly ($\xi = 1.82$) and T4 Lysozyme in H_2O-d_8Gly ($\xi = 1.23$). Going beyond a 5'-DEER setup with higher order CP-type refocusing sequences appears only beneficial at longer time scales. For interpulse delays at which the 5-DEER magnetisation has already decayed, 9-DEER pulse schemes still show 20% of the original 4-DEER echo intensity for both TEMPOL and T4 Lysozyme. The 9-pulse Uhrig scheme allows for even stronger suppression of coherence loss at these longer interpulse delays, provided that ξ is sufficiently greater 1.

Note, how the echo intensity does not decay smoothly but instead shows irregularities in particular for the 7- and 9-pulse schemes at shorter relaxation time. These irregularities are due to contributions from an increasing number of crossing and stimulated echos at the acquisition position upon addition of π refocusing pulses. Small time shifts in the pulse sequence did not alleviate this complication completely. Secondly, higher order refocusing lead to a stronger nuclear modulation of the recorded traces, an effect which is noticable at Q band but dominates X-band relaxation traces. Fig. 13b for T4 Lysozyme in H_2O-d_8Gly and Fig. 14b for MSA in d_8-OTP illustrate this observation. The strong nuclear modulation does not allow for normalisation of the recorded echo intensities to the 4-DEER signal as done otherwise, therefore the axes display absolute echo integrals. The effect is discussed in more detail in section 3.1.6 for MSA.

3.1.5 MS 107

The relaxation behaviour of the rigid monoradical MS 107 at Q Band is shown in Fig. 14a. The echo intensity for the 4-DEER sequence has decayed after roughly 30 µs whereas the 5'-DEER sequence still shows signal intensity up to about twice the time. Out of all higher order decoupling setups, the 9-Uhrig experiment performs best with 20% echo amplitude relative to the initial 4-DEER signal at up to 50 µs. Again, irregularities appear for short interpulse delays. However, as the relaxation measurements for MS 107 were carried out to approximate the relaxation behaviour of the biradical sample MSA 236 for later DEER experiments, relative echo intensities for longer relaxation times are only of interest. These unwanted echo contributions do not contribute for longer interpulse delays.

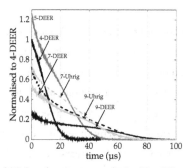

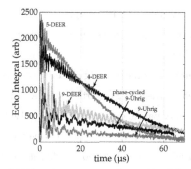

(a) Echo relaxation traces obtained by different pulse sequences 4-DEER (▬), 5'-DEER (▬), 7-DEER ( ), 9-DEER (▬), 7-Uhrig ( ) and 9-Uhrig (▪ ▪) for 50 μM rigid monoradical MS 107 in d_8-OTP at Q band. The echo intensities are normalised to the 4-DEER signal.

(b) Echo relaxation traces obtained by different pulse sequences 4-DEER (▬), 5-DEER (▬), 9-DEER ( ), 9-Uhrig (▬) and 9-Uhrig (▬) with 32-step phase cycle for 50 μM rigid biradical MSA 236 in d_8-OTP at X band. The echo intensites are not normalised due to the strong nuclear modulation visible.

Fig. 14: Relaxation traces for mono- and biradical MS 107 and MSA 236 in d_8-OTP at Q- and X band respectively.

3.1.6   MSA 236

Additionally, the relaxation behaviour of MSA 236 at X band was assessed prior to the corresponding DEER measurements. Fig. 14b shows the echo relaxation traces for a 4-, 5, 9-DEER and 9-Uhrig setup in absolute echo intensities. Comparing the relaxation behaviour to the MS 107 measurements at Q band (Fig. 14a), the relaxation is prolonged for 4- and 5-DEER setups whereas the 9-DEER intensity is comparable for the recorded time window. However, the relative gain for the asymmetric 9-Uhrig scheme as shown for Q band cannot be reproduced for X band. Instead the 9-Uhrig echo intensity is already significantly reduced for small time increments. Overall, the benefit of using higher order decoupling schemes for MSA 236 in X band applies to dipolar evolution times of less than 20 μs where the 5'-DEER setup shows a relative intensity gain. For extreme time scales of above 70 μs the 9-DEER echo intensity becomes only comparable to the standard 4-DEER setup.

The X-band signal shows stronger nuclear modulation. From the 4- and 5'-DEER traces a modulation of 1.1 and 1.25 MHz can be extracted, which can be assigned to matrix deuterons. Yet the 9-Uhrig modulation shows more than one frequency component. This effect arises due to partial excitation of forbidden nuclear transition, similiar to the problem of partial excitation in multipulse DEER sequences discussed in section 2.5.1. The different frequency components overlap and lead to destructive interference, so that the "true" relaxation behaviour is disguised and the echo intensity appears reduced. This effect appears more pronounced for an asymmetric Uhrig-setup than for CP decoupling. In comparison to the relaxation measurement at Q band, the X-band experiment was performed on the biradical instead of the monoradical which implies additional dipolar modulation. Note that the irregularities appearing for short interpulse delays, already described for the Q-band traces, are especially distinct for 9-DEER at X band.

Overall, it is apparent that the depth of the nuclear modulation increases with increasing order n of the employed DD scheme. This trend has been described elsewhere[30] and is significantly stronger in X- relative to Q-band frequencies. Each additional π refocusing pulse redistributes the electron spin coherence also among forbidden transitions. This implies that higher number of pulses do not only refocus electron spin coherence, but also enhance the nuclear modulation effect.

Due to the strong reduction in echo intensity for the 9-Uhrig setup upon going from Q band to X band, the situation at the acquisition position was more closely inspected. It was found that a 32-step phase cycle on [(+x), (-x)] on all five pulses roughly doubles the observed echo intensity as shown in Fig. 14b. This indicates that a negative stimulated echo contribution at the acquisition position cannot be entirely excluded by time shifts.

3.2 DEER experiments

Based on the relaxation characterisation, the DD schemes were extended to corresponding DD-double electron electron resonance (DEER) experiments by adding pump pulses. CP-type DD observer pulse sequences of order order n = 1, 2 and 4 were used to record DEER traces for a rigid biradical model system and T4 Lysozyme in H_2O-d_8-Glycerol. Signal artifacts related to m times pump spin inversion with $1 \leq m < n$ in DD-DEER experiments of order n = 2 or 4 were strongly reduced by increasing pump pulse inversion efficiency. Linear-chirp and asymmetric hyperbolic secant parameters were obtained from pulse optimisation including resonator profile compensation. Extracting the maximal inversion I from $M_z(f)$ measurements, allowed to quantify the inversion efficiency. Experiments at Q band showed pump pulse underperformance with respect to expected inversion efficiencies based on prior Q_{min} calculations. This resulted in strong signal artifact contributions to the corresponding DD-DEER traces. Hence, visible DD-DEER experiments were limited to order n = 2. This excludes Uhrig-type DD schemes which had shown promising echo intensities for large interpulse delays in the respective relaxation study.

At X-band frequencies a nearby full suppression of p_3q_5 signal contributions was achieved, yet with sacrifices in sensitivity relative to Q band. X-band DD-DEER traces of order n = 2 for both MSA 236 and T4 Lysozyme allowed for direct fitting in DeerAnalysis without the need for further data processing. In contrast, signal artifacts at Q-band frequencies led to a slight distortion of the distance distribution upon DeerAnalysis fitting, as the corresponding form factor cannot be matched by the simulated dipolar evolution function. This result calls for an adjustment of the DeerAnalysis kernel for 5'-DEER signal treatment measured at Q-band frequencies.

All displayed DEER signals are normalised to their maximum intensity. The number of traces over which each DEER signal was averaged is specifed in the corresponding figure caption by (number of scans · number of shots per point · phase cycling steps).

3.2.1 MSA 236

The model system MSA 236 has a narrow and well defined distance distribution, as shown in Fig. 15d, and was used to quantify the unwanted signal contribution from incomplete spin inversion at the pump frequency for 5′-DEER and 9′-DEER experiments.

3.2.1.1 Q Band

5′-DEER experiments at Q-Band were carried out with standard 12 ns rectangular pump pulses ("rect") and 200 ns chirp pulses ("chirp") characterised by a rising time of $t_r = 30\text{ns}$, $\Delta\nu = 150$ MHz and starting at ν_{initial} above the observer frequency.

Asymmetric HSh pump pulses are expected to achieve superior inversion characterisation relative to chirp pulses.[11] This could not be confirmed experimentally at Q band for MSA 236 (data not shown), as the unwanted p_3q_5 signal contribution could not be further decreased relative to results for chirp-5′-DEER experiments. The implied limitation in inversion efficiency may be related to B_1 inhomogeneities over the 3 mm diameter sample. In order to test this hypothesis, the oversized resonator was replaced by a commercial Bruker resonator in combination with a reduced sample volume for 1.6 mm sample tubes. The obtained 5′-DEER signal is displayed in Fig. 16 together with traces measured with the oversized resonator. For the latter experimental setup two different modes where chosen to study the influence of the resonator mode choice on the inversion performance. The resonator profiles characterising the three different modes are shown in the bottom half of Fig. 16.

Comparing the p_3q_5 signal contributions for mode 1 and 2 traces in Fig. 16, relates the reduced signal artifact for mode 2 to the choice of a good resonanator mode to allow

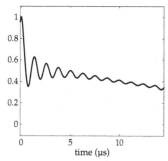

(a) Primary rect-4-DEER data $V(t)$ of 50 µM MSA 236 in d_8-OTP measured at Q band with (11·50·2) scans.

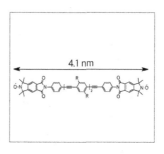

(b) Chemical structure of rigid biradical MSA 236.

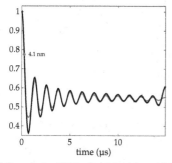

(c) Form factor $F(t)$ (▬) obtained from (a) and a simulated dipolar evolution function $S(t)$(▬).

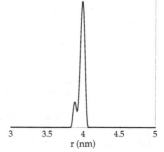

(d) Narrow and well defined distance distribution $P(r)$ obtained by Tikhonov regularisation for a regularisation parameter of α = 0.01.

Fig. 15: 4-DEER signal analysis for the model sample MSA 236 using DeerAnalysis2015.

for sufficient pump pulse bandwidth. The use of the commerical Bruker resonator led to an equally big modulation depth as mode 2 but shows a p_3q_5 signal contribution comparable to mode 1. Overall, mode 2 corresponds to the optimal experimental setup within the limitation posed by the resonator, and was hence used for rect- and chirp-5′-DEER experiments. Fig. 17a and Fig. 17c show the resulting traces together with the reference rect-4-DEER signal. The 4-DEER signal allows to identify the p_3q_5 signal

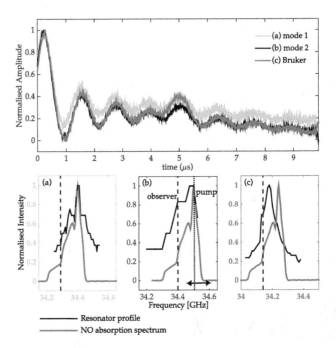

Resonator profile

NO absorption spectrum

Fig. 16: MSA 236 chirp-5′-DEER traces obtained at Q-Band with 200 ns chirps (t_r = 30 ns, $\nu_{initial}$ = 50 MHz and $\Delta\nu$ = 150 MHz) for three different modes all corresponding to (1·50·2) averages. Mode 1 and 2 were measured using an oversized resonator whereas "Bruker" refers to the commerical resonator. The bottom half shows the corresponding resonator profiles together with the NO absorption spectra. For the mode 2 measurements observer and pump position are indicated with the vertical arrow refering to the chirp sweep width.

contribution in equation (30) for the 5′-DEER trace for time $t = \tau$ as depicted in Fig. 3. Both signals were normalised to their maximum intensity and the symmetric 4-DEER signal was fitted to the residual 4-DEER signal contribution as shown in Fig. 17b. This allows for an estimation of $p_3q_5 \approx 0.56$. Upon using chirp pulses for the pump sequence the p_3q_5 contribution can be strongly reduced to 0.17 as shown in Fig. 17c and Fig. 17d. Characterising the efficiency of used pump pulses in terms of the maximum value of inversion I, yields 0.80 for the rectangular shape and 0.90 for the chirped pump pulse. Hence, the inversion I can be related to the reduction in p_3q_5 signal contribution for the 5'-DEER traces. The observed relation between p_3q_5 factor and I supports the argumentation by Spindler and Prisner[5] for using adiabatic inversion pulses to reduce symmetric 4-DEER artifacts.

Note that the frequency of the symmetric 4-DEER signal fits nicely to the residual p_3q_5 signal contribution for Fig. 17b, whereas a noticable phase shift is visible in Fig. 17d. This phase shift is already apparent in Fig. 17c and arises from relative phase gain for spin packets upon using this specific chirp pump pulse with respect to a monochromatic rectuangular pump pulse. By varying the chirp pulse parameters systematically, the relation between the observed phase shift and the employed chirp pulse could potentially be investigated in more detail.

Additionally, the residual signal in Fig. 17d shows a slow baseline rise with increasing time. Besides the phase shift, this reflects the different decaying behaviour of the total 4-DEER and 5'-DEER signal as shown in Fig. 17c. Whereas the 4-DEER trace decays monotonously the 5'-DEER signal shows an increasing and decreasing component towards $t = \tau = 5$ µs due to the superposition of the p_3p_5 and p_3q_5 signal.

The use of chirp pulses achieved a significant reduction in signal artifact, yet an expected improvement for asymmetric HSh pump pulses could not be achieved in Q band. For this a Q band resonator better adopted to this experiment is required. To demonstrate the superior inversion efficiency of HSh pump pulses, additional experiments at lower X-band frequencies were carried out despite the sacrifice in sensitivity.

3.2.1.2   X band

At X band the shaped pulses were chosen shorter in comparison to Q band resulting in 12 ns rectangular pulses ("rect") and 100 ns long chirped pulses ("chirp") with a rise time of $t_r = 25$ ns and with a sweep width of $\Delta\nu = 200$ MHz starting at $\nu_{initial} = 50$ MHz above the chosen observer frequency. Additionally, 100 ns long asymmetric HS pulses of order $h_l = 6$ and $h_r = 1$ were used ("HSh") with $\beta_{HS} = 10.6$ corresponding to a theoretical

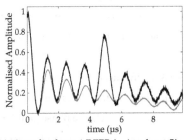

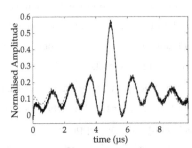

(a) Normalised rect-4-DEER (—) and rect-5'-DEER signal (—). Measured with (2·50·2) and (1·50·2) respectively.

(b) Residual symmetric 4-DEER signal obtained from subtracting 4-DEER from 5'-DEER as shown in Fig. 17a (—) and normalised symmetric rect-4-DEER signal (- -) fitted to (—) which results in an amplitude of 0.56.

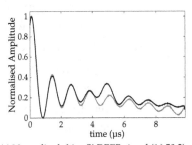

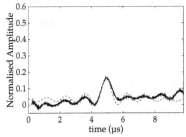

(c) Normalised chirp-5'-DEER signal (16·50·2) (—) obtained with (16·50·2) scans and rect-4-DEER (—).

(d) Residual symmetric 4-DEER signal obtained from subtracting 4-DEER from 5'-DEER as shown in Fig. 17c (—). Normalised symmetric 4-DEER signal measured with 12 ns rectangular pump pulse (- -) fitted to (—) which results in an amplitude of 0.17.

Fig. 17: MSA 236 4- and 5'-DEER traces measured at Q-Band with the mode 2 setup specified in Fig. 16. For the respective pump sequence either 12 ns rectangular pulses ("rect") or 200 ns chirp (t_r = 30 ns, $\nu_{initial}$ = 50 MHz and $\Delta\nu$ = 150 MHz) ("chirp") pulses were used.

optimum.[11]

Fig. 18a illustrates the reduction of the p_3q_5 artifact in 5'-DEER traces by going from

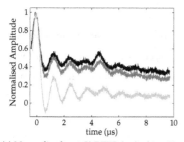

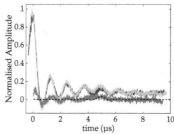

(a) Normalised rect-5'-DEER (▬), chirp-5'-DEER (▬) and HSh-5'-DEER ( ) each obtained with (1·100·8) scans. A reduction in p_3q_5 signal contribution as well as a strong increase in modulation depth is visible for HSh pump pulse.

(b) Normalised HSh-5'-DEER ( ) with HSh-4-DEER (▬) reference signal both corresponding to (6·100·8) averages. The subtraction of HSh-4-DEER from the HSh-5'-DEER trace yields the residual signal (▬). Due to a phase shift the original DEER frequency as indicated by symmetric HSh-4-DEER signal (▪ ▪) is not recovered but an estimate of $p_3q_5 \approx 0.036$ can be obtained.

Fig. 18: 5'-DEER and reference 4-DEER traces for 50 μM rigid biradical MSA 236 in d_8-OTP at X band. The p_3q_5 signal contribution is assessed depending on the applied pump pulse shape.

rectangular over chirp to HSh pump pulses. The latter shows a strong increase in modulation depth additionally to a near complete suppression of the 4-pulse artifact. A DEER-type modulation was also observed in the imaginary part of the signal for chirp and even more strongly for HSh pump pulses.

Fig. 18b compares the HSh-5'-DEER signal to the reference HSh-4-DEER trace, which is very similar in its progess in time,§ yet a relative phase shift is visible. Upon subtraction of the two normalised signals, the symmetric 4-DEER trace is expected. Within experimental uncertainty, it is difficult to identify the shape of the residual 4-DEER signal. Therefore the symmetric 4-DEER trace was only used to estimate p_3q_5 by scaling the symmetric 4-DEER signal with a factor of 0.036. This result shows that the HSh pulse shape is superior to the linear chirp pulses used here.

3.2.1.3 DEER data analysis

The strong reduction in artifical signal contribution allowed to test in how far the residual artifact distorts the computation of the underlying distance distribution $P(r)$ with DeerAnalysis.[28] The results for the unmodified HSh-5'-DEER signal are shown in Fig. 19 for α values 0.1, 1 and 10.

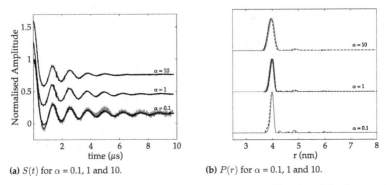

(a) $S(t)$ for α = 0.1, 1 and 10. (b) $P(r)$ for α = 0.1, 1 and 10.

Fig. 19: Comparison between the dipolar evolution function $S(t)$ (▬) and the distance distributions $P(r)$ obtained by Tikhonov regularisation for different regularisation parameters α. Reference HSh-4-DEER trace (▬ ▬) and uncorrected HSh-5'-DEER signal (▬) for 50 µM MSA 236 in d$_8$-OTP at X band obtained with (6·100·8) averages.

The increasingly bigger regularisation parameter α introduces artificial broadening of the distance distribution such that for α = 10 distance distributions $P(r)$ obtained from HSh-4-DEER virtually coincide with the corresponding HSh-5'-DEER $P(r)$ as shown in Fig. 19b. Going to smaller α values, the deviation is small, yet affects less certain distance contributions at about 6 nm more strongly. Overall, these results indicate that the use of asymmetric HSh pump pulses allows to reduce p_3q_5 artifacts to an extent that the HSh-5'-DEER signal can be directly used for DEER data analysis without the need to account for the 4-DEER contribution. So far the benefit of 5'-DEER experiments was mainly achieved by subtraction of an additionally recorded symmetric 4-DEER trace prior to further data analysis.[4] Alternatively, the default kernel in DeerAnalysis for the

4-DEER setup would need to be adopted to the 5'-DEER signal contributions defined by equation (30). For the MSA 236 probe this analytically more rigorous approach for 5'-DEER analysis is not required as the high number of oscillations allows for a good fit in terms of frequency which translates into a well defined distance distribution. Owing to higher inversion efficiencies at X band the extension of the 5'-DEER experiment to a 9'-DEER seemed feasible. However, the results from X-band relaxation measurements, shown in Fig. 14 in page 51, allow to identify only half of the echo intensity at $t = 20$ μs for a 9'-DEER setup relative to a 4 or 5'-DEER experiment.

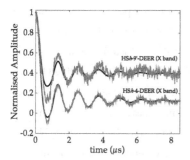

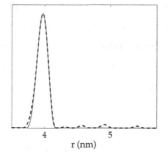

(a) Background corrected form factor $F(t)$ for HSh-9'-DEER and HSh-4-DEER (—) and $S(t)$ (—). The 9'-DEER trace shows a reduced modulation depth and cannot be fitted by $S(t)$ due to the underlying 4-DEER kernel.

(b) Comparison between distance distributions $P(r)$ obtained by Tikhonov regularisation for $\alpha = 1$ from the respective $S(t)$ in Fig. 20a. Reference $P(r)$ obtained from HSh-4-DEER (▪ ▪) and $P(r)$ obtained from HSh-9-DEER (—) with an average distance of 4.0 and 3.8 nm respectively.

Fig. 20: MSA 236 asymmetric HSh-9'-DEER trace in comparison to a HSh-4-DEER trace measured at X band corresponding to (6·100·8) and (7·100·8) averages respectively. With a pump pulse parameter for "HSh" being $t_p = 100$ ns, $\nu_{initial} = 50$ MHz, $\Delta\nu = 200$ MHz and $\beta_{HS} = 10.4$.

Fig. 20a shows the 9'-DEER trace relative to a standard 4-DEER setup. The decrease in signal-to-noise (S/N) ratio as indicated by the X-band relaxation measurements is noticable. Moreover, the modulation depth of the 9-pulse signal appears reduced. This effect is expected as the principle modulation by $\cos[2d(t - \delta T)]$ scales with p^4 relative to a weight of p for the 4-DEER modulation $\cos[d(t_1 - t)]$ for 4-DEER time delays as shown in Fig. 3a on page 15.

The simulated dipolar evolution function $S(t)$ matches the 4-DEER trace well while deviating for the 9'-DEER signal. Due to the high number of oscillations the fit still obtains the underlying dipolar frequency which leads to a similiar distance distribution as derived from the 4-DEER signal (Fig. 20b).

Equation (37) shows all the expected contributions to the measured 9-pulse signal where only dipolar pathways with weights down to $p^3(1-p)$ have been considered. In addition to the principal 9'-DEER modulation with weight p^4 and zero time $t = \delta T$, a second contribution refocused at $t = \delta T/2$ weighted by $p^3(1-p)$ is expected. This explains the mismatch between the form factor $F(t)$ and $S(t)$ as the latter is derived for a 4-DEER kernel. Signal contributions which are expected at $t \approx \tau/4$ leading to a symmetric signal artifact, oscillating with half the dipolar frequency, cannot be identified by eye. Note that the background decreases monotonically, indicating a small symmetric signal contribution. Yet the reduced sensitivity at X band complicates a rigorous signal analysis with respect to possible contributions for a 9-pulse setup at the here shown S/N level. Residual signal contribution could potentially be identified upon increasing the number of averages at X band. Alternatively, a higher S/N level would be expected at Q band which can be approximated by a third of the 5'-DEER S/N ratio. This estimate can be derived from Fig. 14 on page 51 for $t = 20$ μs. Yet attempts to measure a 9'-DEER signal at these higher frequencies failed due to insufficient inversion efficiency.

3.2.2 T4 Lysozyme in H_2O-d_8Gly

The model system MSA 236 allowed for 5'- and 9'-DEER measurements in which the benefit of HSh inversion efficiencies were showcased and reduction in unwanted signal contributions was achieved. Data analysis with the standard DeerAnalysis programme allowed to obtain a distance distribution matching $P(r)$ derived from 4-DEER signals without 5'- and 9'-DEER data post processing or adjustment of the kernel. For samples with a restricted number of measurable dipolar periods this straight-forward data analysis may break down. Therefore T4 Lysozyme was chosen as a sample to mimic a biologically relevant spin environment with shorter T_m times. This limits the measurable number of dipolar periods.

The relaxation behaviour of T4 Lysozyme in H_2O-d_8Gly can be characterised by T_m = 2.60 and ξ = 1.23 from 4-DEER relaxation measurements in Q band. The traces obtained at Q- and X-band frequencies are shown in Fig. 13a and Fig. 13b on page 49, respectively. The latter were neither normalised nor fitted for reasons already discussed in section 3.1.6.

The low spin concentration of 13 μM was chosen to minimise the background decay present in the DEER signal. This allows to treat the signal as $V(t) \approx F(t)$, neglecting the background contribution $B(t)$ to first approximation.

3.2.2.1 Q band

The Q-band relaxation traces shown in Fig. 13a on page 49 illustrate the gain of a 5-pulse setup relative to the 4-DEER relaxation measurement as expected for $\xi > 1$. At $t > 20$ μs 9-DEER observer sequences allow for detectable echo intensity, whereas signals from a 4- and 5 pulse setup have already decayed. Already at $t > 10$μs show 7- and 9-Uhrig schemes higher signal intensity relative to a 5-DEER setup.

Pump pulse optimisation for the T4 Lysozyme sample achieved a maximal I value of 0.91 for both chirped and asymmetric HSh pump pulses, whereas the latter showed a flatter inversion maximum as illustrated in Fig. 21. Hence, the HSh was used to record 5'-DEER traces as shown in Fig. 21. The optimal pump pulse length t_p of 200 ns to suppress the 4-DEER signal contribution was experimentally determined.

Fig. 22 compares the rect-4-, asymmetric HSh-4- and 5'-DEER traces which all display a different modulation depth Δ. In an attempt to increase the inversion efficiency of the HSh pump pulse further, the sample volume was reduced to two thirds of the orginal 38 μL sample volume. This allowed for 14 ns long π-pulses instead of a pulse length of 20

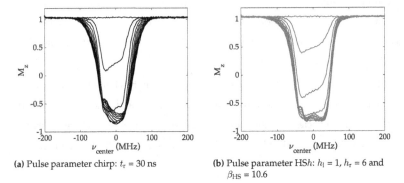

(a) Pulse parameter chirp: t_r = 30 ns

(b) Pulse parameter HSh: h_l = 1, h_r = 6 and β_{HS} = 10.6

Fig. 21: Longitudinal magnetisation S$_z$ for 200 ns long and 100 MHz broad chirp (—) and HSh (—) pulse for increasing pulse amplitudes 0 to 1 for $\nu_{initial}$ = 50 MHz. The HSh pulse appears with a flatter inversion maximum. Both pulses only achieve $I \approx 91\%$ at Q band.

ns for the 38 μL sample volume. Volume reduction also led to an increased modulation depth, indicating that a bigger fraction of spins were excited and contributed the the signal. Therefore, reduction of B_1 inhomogeneity indead appears to be one of the experimental imperfections causing reduced pump pulse performance in Q band. Upon subtraction of the HSh-4- and 5'-DEER traces as done in Fig. 22b, the residual signal displays a period stretched by factor 1.2 relative to the reference 4-DEER dipolar period. The factor of 1.2 reduces to 1.1 for X-band HSh measurements where the used chirp had a narrower bandwidth.

Fig. 22a furthermore illustrates the increased modulation depth when using HSh relative to rectangular pump pulses which indicates that a larger number of spins contribute to the signal.

3.2.2.2 X band

Fig. 13b on Fig. 13b depicts the relaxation behaviour of T4 Lysozyme at X band and indicates that DEER measurement of DD order $n \geq 3$ are not feasible due to low echo

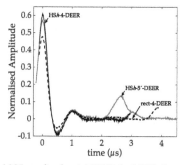
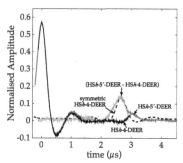

(a) Normalised rect-4-DEER (- -), HS*h*-4-DEER (—) and HS*h*-5'-DEER (—) traces. The 4-DEER signal modulation depths Δ are 0.4775, 0.6097 and 0.5726 for the respective traces. All signals were obtained with (174·50·2) averages.

(b) Normalised HS*h*-5'-DEER (—) with scaled HS*h*-4-DEER (—) to match the HS*h*-5' modulation depth. The subtraction of HS*h*-4-DEER from HS*h*-5'-DEER trace yields the residual signal (—). Due to a phase shift the original DEER frequency as indicated by the symmetric HS*h*-4-DEER (- -) is not recovered. Still, p_3q_5 can be approximated to 0.138.

Fig. 22: T4 Lysozyme in H_2O-d_8Gly 4- and 5'-DEER traces in Q-Band. For the respective pump sequence a 12 ns rectangular pulses ("rect") or 200 ns HS*h* ($\nu_{initial}$ = 50MHz, $\Delta\nu$ = 100 MHz, h_l = 1, h_r = 6, amplitude scale = 0.5, β_{HS} = 10.6) ("HS*h*") pulses were used.

intensities. In comparison to Q band, compare Fig. 13a, a 9-Uhrig setup is not superior to the 5-DEER experiment for longer interpulse delays.

Note that DEER measurements at X band are usually performed at higher concentration, whereas the sampled employed here was optimised for Q-band measurements. 4- and 5'-DEER experiments with HS-pump pulse (h_l = 6, h_r = 1, t_p = 100 ns, $\nu_{initial}$ = 50 MHz, $\Delta\nu$ = 200 MHz and β_{HS} = 10.4) as shown in Fig. 23 were measured. Fig. 23 shows the improved inversion efficiency in X band for asymmetric HS*h* pulses where 4 and 5'-DEER signal overlap within the limits of the S/N level. The arrow indicates where the symmetric 4-DEER signal would be expected.

Comparing the 5-DEER signal in X band (Fig. 23) to Q-band data (Fig. 22a), illustrates the reduced sensitivity at X-band frequencies. Even though Q-band data was acquired

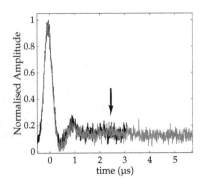

Fig. 23: Normalised HS*h*-4 (▬) and 5′-DEER (▬) traces for T4 Lysozyme in H$_2$O-d$_8$Gly in X band with (14·100·8) averages. The asymmetric HS*h*-pulse is characerised by $h_1 = 6$, $h_r = 1$, $t_p = 100$ ns, $\nu_{initial} = 50$ MHz, $\Delta\nu = 200$ MHz and $\beta_{HS} = 10.4$ for the first pump pulse in both sequences. For the second pump pulse in the 5′ setup the order and sweep direction were reversed. The arrow indicates the time t at which the 4-DEER artifact would be expected.

with a 1.5-fold number of averages relative to X band, the S/N ratio has decreased by more than a factor of $\frac{(S/N)_{X\,band}}{(S/N)_{Q\,band}} = 0.81$. However, the modulation depth visible in Fig. 23 is notably large. This implies that less than 10 % of spins serve as observer spins which limits the detectable signal.

3.2.2.3 DEER data analysis

Fig. 24 compares the distance distributions derived from HS*h*-5′-DEER signals both at X and Q band. A rect-4-DEER trace measured at Q band serves as a reference. The corresponding distance distributions $P(r)$ are displayed in Fig. 24b for a regularisation parameter $\alpha = 100$. The 4-DEER reference data yields a distance distribution charac-terised by a single peak at 3.5 nm. Upon analysis of the X-band signal the same average distance of 3.5 nm between spin position 71 and 131 in the double mutant T4 Lysozyme is obtained. However, the distribution includes an artifact at roughly 5.1 nm. This second peak is even more pronounced for $P(r)$ derived from the 5′-DEER Q-band signal. Fig. 24a indicates where this arifact arises from. The simulated dipolar evolution function $S(t)$ interprets the $p_3 q_5$ signal contribution as a dipolar frequency component. $S(t)$ for the Q-band 5'-DEER trace shows a notiable vertical extension exactly where the arrow in Fig. 23 points. The signal artifact leads not only to an additional peak at larger distances, but slightly distorts the right shoulder of the true distance distribution known from the reference data. This distortion originates from the 4-DEER signal contribution creating an artifical bump at 1.7 μs. Overall, the HSh-5'-DEER Q-band data yields an overestimation of the average distance by 0.1 nm. On the other hand, it illustrates nicely how the dipolar evolution time is extended by 2 μs for identical S/N level and number of scans at Q band. Secondly, the latter traces allows to identify the individual signal contributions by eye. In contrast, at X band only the artifical peak at 5.1 nm in the corresponding $P(r)$ hints towards a 4-DEER signal impurity at 1.7 μs. Otherwise, $S(t)$ appears to fit the X-band data nicely in the limits of the S/N level as visible in Fig. 24a.

3.2.3 Comparison X- and Q-band DEER experiments

The results obtained for MSA 236 and T4 Lysozyme illustrate the tradeoff between sensitivity and inversion efficiency at Q- and X-band frequencies. These findings imply that the issue of symmetric 4-DEER contributions in the 5'-DEER signal may be either solved spectroscopically or via data post-processing. The latter approach was introduced by Borbat and Freed[4]. Yet they discuss "the room for improvement" in the corresponding SI using instrumental efforts and state that "pure shaped pulses ... (modified Hermite and hyperbolic secant types) did not provide major improvement compared to a 12 ns rectangular pulse". However they acknowledge that upon further modification the inversion efficiency would improve substantially. This has been showcased in this thesis as shaped asymmetric HSh pump pulses at X band achieved virtual suppression of the symmetric 4-DEER artifacts. Consequently, the 5'-DEER was extended to a 9'-DEER setup in case of MSA 236. However, the measurements at X-band frequencies lead to a decrease in S/N level due to the reduced sensitivity with respect to Q band.

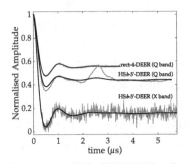
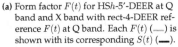
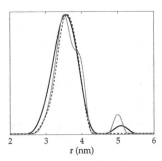

(a) Form factor $F(t)$ for HSh-5'-DEER at Q band and X band with rect-4-DEER reference $F(t)$ at Q band. Each $F(t)$ (—) is shown with its corresponding $S(t)$ (—).

(b) Comparison between distance distributions $P(r)$ obtained by Tikhonov regularisation for $\alpha = 100$ from the respective $S(t)$ in Fig. 24a. Rect-4-DEER data at Q band (- -) served as a reference for the HSh-5'-DEER distance distributions at X (—) and Q band (—).

Fig. 24: Comparison of DeerAnalysis fitting for T4 Lysozyme traces in H_2O-d_8Gly with regularisation parameter $\alpha = 100$. With reference rect-4-DEER signal at Q band and HSh-5'-DEER traces measured at Q- and X-band frequencies. The HSh-5'-DEER traces are correspond to DEER data shown in Fig. 22 and Fig. 23 respectively. The rect-4-DEER reference signal was measured at Q band with (174·50·2) averages.

At Q band, 4-DEER signal suppression has been shown to be sensitive to B_1 inhomogeneities as well as the choice of the resonator mode. Yet, with the current experimental setup further reduction of signal artifacts at Q-band frequencies were not possible. Instead of spectectroscopic efforts, data post-processing is an option to benefit from the advantages of 5'-DEER experiments. The standard data analysis procedure for the 4-DEER setup is described in section 1.4.4. Extracting the underlying distance distribution from a 5'-DEER signal with symmetric 4-DEER artifacts hence requires an adjustment of the background signal $B(t)$ treatment as well as of the Tikhonov kernel $K(t,r)$.

4 Conclusion and Outlook

The relaxation behaviour of model systems was characterised for both CP and Uhrig-type pulse sequences of constant sequence length. Despite better suppression of coherence losses for an Uhrig setup in Q-band relaxation measurements, this advantage did not translate into corresponding DEER measurements as 9-pulse DEER experiments were not realisable due to pump pulse inversion inefficiency. At X band signal loss due to destructive interference of frequencies introduced by nuclear modulation leads to an overall better performance of CP-type decoupling with respect to asymmetric Uhrig-type schemes.

Hence, CP-type DD DEER experiments of order $n=2$ and 4 were recorded in combination with optimised pump pulse design, allowing for a significant suppression of signal artifacts arising from an incomplete pump pulse inversion, in particular at X Band. Signal artifacts arising at Q band for identical pump pulse choice are due to experimental imperfects which are not completely understood yet.

X-band DD-DEER traces of order $n = 2$ for the rigid biradical sample allowed for direct DeerAnalysis fitting without the need for data post processing. Therefore the MSA 236 sample illustrates a case in which the 4-DEER signal contribution was dealt with spectroscopically. This corresponds to a procedure which is desirable also for samples in which the respective DEER signal displays a smaller number of dipolar periods guiding the fit. This is the case for T4 Lysozyme. In the low concentration regime, where the background contribution is negligible, there is currently a choice between higher sensitivity and less signal artifact. This was shown for T4 Lysozyme sample, for which DEER traces were recorded both at Q- and X-band frequencies. Whereas the latter allows for better pump pulse inversion, the individual signal fractions are more difficult to be identified due to a lower S/N ratio. In contrast, the Q-band 5'-DEER signal allows for identification of both p_3q_5- and p_3p_5 signal contributions. This corresponds to a case where the 4-DEER signal artifact can be separated analytically by extending the default DeerAnalysis kernel, so that both signal contributions can be fitted.

In this thesis the maximum value of the inversion efficiency I was used to characterise the pump pulse performance and was related to the 4-DEER signal artifact amplitude. More rigorously, the inversion quality κ, as defined in equation (26), can be derived from experiment to determine the expected weights p_3p_5 and p_3q_5 for both 5'- and 4-DEER signal contribution, respectively. This may serve as a starting point for a least square fit to the DEER traces with a 5'-DEER adjusted kernel. Such a procedure would allow to treat current 5'-DEER data at Q-Band frequencies.

In order to assess the stability of this approach simulated 5'-DEER traces with known signal contribution weights have to be generated and the fit based on the adjusted kernel has to be tested. This approach can be further extended to higher order 9'-DEER schemes based on equation (37), provided that S/N level allows for the identification of signal components. Secondly, cases in which a background decay is present need to be studied, to extend the procedure to typical DEER applications. For this case higher concentrated samples, for instance of T4 Lysozyme, need to be prepared. This allows to determine experimentally how the background function is composed in a 5'- or 9'-DEER experiment. For the moment, the same background decay behaviour for "pure" 4- and 5-DEER signal is expected, as the observer sequence is identical. Hence, the total 5'-DEER background signal is a superposition of both contributions, with known decay starting points at times, when the corresponding dipolar pathway is refocused. The total background signal is expected to become more complicated for a 9'-DEER signal. Only if the correct functional form of the background contribution is known, the signal separation in background and form factor can be carried out.

It is spectroscopically desirable to improve pump pulse inversion efficiency at Q-Band frequencies. Apart from a sensitivity increase, this would also allow for an Uhrig-type

9'-DEER setup. This scheme promises a superior signal intensity for sufficient high ξ values at longer interpulse delays with respect the CP-derived DEER experiment of order $n = 4$. Furthermore, DEER experiments at Q band are less dominated by nuclear modulation as it is currently the case for X-band DEER signals. The latter can be alternatively suppressed by variation of the fixed interpulse delays and summing over the different delays times as it is done in 4-DEER experiments.

This thesis has affirmed that dynamical decoupling can be used to extend the standard 4-DEER experiment, allowing for longer dipolar evolution times. However, the order n of the decoupling scheme which is feasible for an experimental DEER setup is limited due to the increasing number of dipolar pathways and complications arising from imperfect pump and observer pulses. For the moment, 5'-DEER experiments with pump pulse optimisation using the described GUI interface represent the biggest improvement for DEER measurements at X band.

Bibliography

References

[1] Huber, M.; Lindgren, M.; Hammarström, P.; Mårtensson, L.-G.; Carlsson, U.; Eaton, G.; Eaton, S. *Biophysical Chemistry* **2001**, *94*, 245–256.

[2] Eaton, S. S.; Eaton, G. *Distance Measurments in Biological Systems by EPR*, Vol. 19; Springer US, 2002.

[3] Slichter, C. P. *Principles of Magnetic Resonance*; Springer-Verlag Berlin Heidelberg New York, second revised and expanded edition ed., 1978.

[4] Borbat, P.; Georgieva, E. R.; J.H., F. *The Journal of Physical Chemistry Letters* **2013**, *4*, 170–175.

[5] Spindler, P. E.; Glaser, S. J.; Skinner, T. E.; Prisner, T. F. *Angew. Chem. Int. Ed.* **2013**, *52*, 3425–3429.

[6] Jeschke, G. Pciv: Magnetic resonance: Electron paramagnetic resonance spectroscopy; December , **2013**.

[7] Schweiger, A.; Jeschke, G. *Principle of pulse electron paramagnetic resonance*; Oxford University Press, 2001.

[8] Tannus, A.; Garwood, M. *Journal of Magnetic Resonance* **1996**, *120*, 133–137.

[9] Baum, J.; Tycko, R.; Pines, A. *Phys. Rev. A* **1985**, *32*, 3435–3447.

[10] Kupce, E.; Freemann, R. *Journal of Magnetic Resonance* **1996**, *118*, 299–303.

[11] Dounas, A. Pulse optimisation for efficient spin inversion Semester project, Eidgenössische Technische Hochschule Zürich, **Spring Semester 2015**.

[12] Pannier, M.; Veit, S.; Godt, A.; Jeschke, G.; Spiess, H. *Journal of Magnetic Resonance* **2000**, *142*(2), 331–340.

[13] Milov, A. D.; Salikhov, K.; Shirov, M. D. *Soviet Physics-Solid State* **1981**, *23*(4), 565–9.

[14] Jeschke, G. *Annual review of physical chemistry* **2012**, *63*, 419–446.

[15] Polyhach, Y.; Bordignon, E.; Tschaggelar, R.; Gandra, S.; Godt, A.; Jeschke, G. *Phys. Chem. Chem. Phys.* **2012**, *14*, 10762–10773.

[16] Mehring, M.; Waugh, J. *Rev. Sci. Instrum.* **1991**, *43*, 649–653.

[17] Kurshev, V. V.; Raitsimring.; Tsvetkov, Y. D. *Journal of Magnetic Resonance* **1989**, *81*(3), 441–454.

[18] Deeranalysis2013.2 user manual. Jeschke, G.; ETH Zürich, August , 2013.

[19] Jeschke, G.; Bender, A.; Paulsen, H.; Zimmermann, H.; Godt, A. *Journal of Magnetic Resonance* **2004**, *169*, 1–12.

[20] Dastvan, R.; Bode, B. E.; Karuppiah, M. P. R.; Marko, A.; Lyubenova, S.; Schwalbe, H.; Prisner, T. F. *J. Phys. Chem. B* **2010**, *114*, 13507–13516.

[21] Jeschke, G.; Polyhach, Y. *Phys. Chem. Chem. Phys.* **2007**, *9*, 1895–1910.

[22] Uhrig, G. *Physical Review Letters* **2007**, *98*.

[23] Lee, B.; Witzel, W. M. S. S. D. *Phys. Rev. Lett* **2008**, *100*.

[24] Doppelbauer, M. Dynamic nuclear polarization and relaxation based water accessibility measurements of spin Semester project, Eidgenössische Technische Hochschule Zürich, **Fall Semester 2014.**

[25] Doll, A.; Jeschke, G. *Journal of Magnetic Resonance* **2014**, *246*, 18–26.

[26] Doll, A.; Qi, M.; Pribitzer, S.; Willi, N.; Yulikov, M.; Godth, A.; Jeschke, G. *Phys. Chem. Chem. Phys.* **2015**, *17*, 7334–7344.

[27] Doll, A.; Pribitzer, S.; Tschaggelar, R.; Jeschke, G. *Journal of Magnetic Resonance* **2013**, *230*, 27–39.

[28] Jeschke, G.; Chechik, V.; Ionita, P.; Godt, A.; Zimmermann, H.; Banham, J. *Appl. Magn. Reson.* **2006**, *30*, 473–498.

[29] Zecevic, A.; Eaton, G. R.; Eaton, S. S.; Lindgren, M. *Molecular Physics* **1998**, *95*(6), 1255–1263.

[30] Mitrikas, G.; Prokopiou, G. *Journal of Magnetic Resonance* **2015**, *254*, 78–85.

Printed in the United States
By Bookmasters